ENERGY SCIENCE, ENGINEERING AND TECHNOLOGY

SHALE GAS DEVELOPMENT

ENERGY SCIENCE, ENGINEERING AND TECHNOLOGY

Additional books in this series can be found on Nova's website at:

https://www.novapublishers.com/catalog/index.php?cPath=23_29&seriesp=Energy+Science%2C+Engineering+and+Technology

Additional e-books in this series can be found on Nova's website at:

https://www.novapublishers.com/catalog/index.php?cPath=23_29&seriespe=Energy+Science%2C+Engineering+and+Technology

ENERGY SCIENCE, ENGINEERING AND TECHNOLOGY

SHALE GAS DEVELOPMENT

KATELYN M. NASH
EDITOR

Nova Science Publishers, Inc.
New York

LIBRARY OF CONGRESS CATALOGING-IN-PUBLICATION DATA

Shale gas development / Editor, Katelyn M. Nash.

ix, 174 p. : ill., maps ; 26 cm.

Includes bibliographical references and index.

ISBN 978-1-61668-545-4 (hardcover)

1. Natural gas --Prospecting --Environmental aspects --United States. 2. Natural gas --Prospecting --Marcellus Shale. 3. Natural gas --Prospecting --Government policy --United States. 4. Natural gas reserves --United States. 5. Marcellus Shale. I. Nash, Katelyn M.

622.33850973

2011281627

Published by Nova Science Publishers, Inc. † New York

CONTENTS

PREFACE

Natural gas production from hydrocarbon-rich shale formation, known as "shale gas", is one of the most rapidly expanding trends in onshore domestic oil and gas exploration and production today. In some areas, this has included bringing drilling and production to regions of the country that have seen little or no activity in the past. New oil and gas developments bring changes to the environmental and socio-economic landscape, particularly in those areas where gas development is a new activity. With these changes have come questions about the nature of shale gas development, the potential environmental impacts, and the ability of the current regulatory structure to deal with this development. This book discusses geologic information on the shale gas basins in the U.S. and the methods of shale gas development, as well as the regulatory framework and the environmental considerations associated with shale gas development.

Chapter 1 - This Primer on Modern Shale Gas Development in the United States was commissioned through the Ground Water Protection Council (GWPC). It is an effort to provide sound technical information on and additional resource development activity, and environmental protection, especially water resource management. The GWPC is the national association of state ground water and underground injection agencies whose mission is to promote the protection and conservation of ground water resources for all beneficial uses. One goal of the GWPC is to provide a forum for stakeholder communication on important current issues to foster development of sound policy and regulation that is based on sound science. This Primer is presented in the spirit of furthering that goal.

Water and energy are two of the most basic needs of society. Our use of each vital resource is reliant on and affects the availability of the other. Water is needed to produce energy and energy is necessary to make water available for use. As our population grows, the demands for both resources will only increase. Smart development of energy resources will identify, consider, and minimize potential impacts to water resources.

Natural gas, particularly shale gas, is an abundant U.S. energy resource that will be vital to meeting future energy demand and to enabling the nation to transition to greater reliance on renewable energy sources.

Shale gas development both requires significant amounts of water and is conducted in proximity to valuable surface and ground water. Hence, it is important to reconcile the concurrent and related demands for local and regional water resources, whether for drinking water, wildlife habitat, recreation, agriculture, industrial or other uses.

Because shale gas development in the United States is occurring in areas that have not previously experienced oil and gas production, the GWPC has recognized a need for credible, factual information on shale gas resources, technologies for developing these resources, the regulatory framework under which development takes place, and the practices used to mitigate potential impacts on the environment and nearby communities. also addresses nonwater issues that may be of interest to citizens, government officials, water supply and use professionals, and other interested parties.

Each state has laws and regulations to ensure the wise use of its natural resources and to protect the environment. The GWPC has conducted a separate study to summarize state oil and gas program requirements that are designed to protect water resources. These two studies complement one other and together provide a body of information that can serve as a basis for fact-based dialogue on how shale gas development can proceed in an environmentally responsible manner under the auspices of state regulatory programs.

This Shale Gas Primer was intended to be an accurate depiction of current factors and does not represent the view of any individual state. Knowledge about shale gas development will continue to evolve. The GWPC welcomes insights that readers may have about the Primer and the relationship of shale gas development to water resources.

Chapter 2 - In the past, the oil and gas industry considered gas locked in tight, impermeable shale uneconomical to produce. However, advances in directional well drilling and reservoir stimulation have dramatically increased gas production from unconventional shales. The United States Geological Survey estimates that 200 trillion cubic feet of natural gas may be technically recoverable from these shales. Recent high natural gas prices have also stimulated interest in developing gas shales. Although natural gas prices fell dramatically in 2009, there is an expectation that the demand for natural gas will increase. Developing these shales comes with some controversy, though.

The hydraulic fracturing treatments used to stimulate gas production from shale have stirred environmental concerns over excessive water consumption, drinking water well contamination, and surface water contamination from both drilling activities and fracturing fluid disposal.

The saline "flowback" water pumped back to the surface after the fracturing process poses a significant environmental management challenge in the Marcellus region. The flowback's high content of total dissolved solids (TDS) and other contaminants must be disposed of or adequately treated before discharged to surface waters. The federal Clean Water Act and state laws regulate the discharge of this flowback water and other drilling wastewater to surface waters, while the Safe Drinking Water Act (SDWA) regulates deep well injection of such wastewater. Hydraulically fractured wells are also subject to various state regulations. Historically, the EPA has not regulated hydraulic fracturing, and the 2005 Energy Policy Act exempted hydraulic fracturing from SDWA regulation. Recently introduced bills would make hydraulic fracturing subject to regulation under SDWA, while another bill would affirm the current regulatory exemption.

Gas shale development takes place on both private and state-owned lands. Royalty rates paid to state and private landowners for shale gas leases range from 12½ % to 20%. The four states (New York, Pennsylvania, Texas, and West Virginia) discussed in this chapter have shown significant increases in the amounts paid as signing bonuses and increases in royalty rates. Although federal lands also overlie gas shale resources, the leasing restrictions and the low resource-potential may diminish development prospects on some federal lands. The practice of severing mineral rights from surface ownership is not unique to the gas shale development. Mineral owners retain the right to access surface property to develop their holdings. Some landowners, however, may not have realized the intrusion that could result from mineral development on their property.

Although a gas-transmission pipeline-network is in place to supply the northeast United States, gas producers would need to construct an extensive network of gathering pipelines and supporting infrastructure to move the gas from the well fields to the transmission pipelines, as is the case for developing any new well field.

Chapter 3 - The Marcellus Shale is a sedimentary rock formation deposited over 350 million years ago in a shallow inland sea located in the eastern United States where the present-day Appalachian Mountains now stand (de Witt and others, 1993). This shale contains significant quantities of natural gas. New developments in drilling technology, along with higher wellhead prices, have made the Marcellus Shale an important natural gas resource.

The Marcellus Shale extends from southern New York across Pennsylvania, and into western Maryland, West Virginia, and eastern Ohio (figure 1). The production of commercial quantities of gas from this shale requires large volumes of water to drill and hydraulically fracture the rock. This water must be recovered from the well and disposed of before the gas can flow. Concerns about the availability of water supplies needed for gas production, and questions about wastewater disposal have been raised by water-resource agencies and citizens throughout the Marcellus Shale gas development region. This Fact Sheet explains the basics of Marcellus Shale gas production, with the intent of helping the reader better understand the framework of the water-resource questions and concerns.

In: Shale Gas Development
Editor: Katelyn M. Nash pp.1-104
ISBN: 978-1-61668-545-4

Chapter 1

MODERN SHALE GAS: DEVELOPMENT IN THE UNITED STATES - A PRIMER*

U.S. Department of Energy
Office of Fossil Energy, National Energy Technology Laboratory

ACKNOWLEDGMENTS

This material is based upon work supported by the U.S. Department of Energy, Office of Fossil Energy, National Energy Technology Laboratory (NETL) under Award Number DE-FG26-04NT15455. Mr. Robert Vagnetti and Ms. Sandra McSurdy, NETL Project Managers, provided oversight and technical guidance. This study was directed by the Ground Water Protection Council (GWPC) with ALL Consulting serving as lead researcher.

GWPC and ALL Consulting wish to extend their appreciation to the following federal, state, industry, and educational institutions which helped with numerous data sources, data collection and technology reviews that were critical to the success of this project. Additionally, the extra time and energy that individuals provided in reviewing and in broadening our understanding of the issues at hand is respectfully acknowledged.

The authors wish to specifically acknowledge the help and support of the following entities: Arkansas Oil and Gas Commission, Louisiana Department of Natural Resources, Michigan Department of Environmental Quality Office of Geological Survey, Montana Board of Oil and Gas Conservation, Montana Department of Natural Resources, New York State Department of Environmental Conservation, Ohio Department of Natural Resources Division of Mineral Resources Management, Oklahoma Corporation Commission, Pennsylvania Department of Environmental Protection, Railroad Commission of Texas, State of Tennessee, State University of New York at Fredonia, West Virginia Department of Environmental Protection, Energy Information Administration, U.S. Environmental Protection Agency, State Review of Oil and Natural Gas Environmental Regulation, Inc. (STRONGER), BP America

* This is an edited, reformatted and augmented version of a U. S. Department of Energy publication dated April 2009.

Production Co., Chesapeake Energy Corp., Devon Energy Corp., East Resources, Inc., Fortuna Energy Inc., Independent Petroleum Association of America, Schlumberger Ltd., Universal Well Services Inc., and Weatherford International Ltd.,

FOREWORD

This Primer on Modern Shale Gas Development in the United States was commissioned through the Ground Water Protection Council (GWPC). It is an effort to provide sound technical information on and additional resource development activity, and environmental protection, especially water resource management. The GWPC is the national association of state ground water and underground injection agencies whose mission is to promote the protection and conservation of ground water resources for all beneficial uses. One goal of the GWPC is to provide a forum for stakeholder communication on important current issues to foster development of sound policy and regulation that is based on sound science. This Primer is presented in the spirit of furthering that goal.

Water and energy are two of the most basic needs of society. Our use of each vital resource is reliant on and affects the availability of the other. Water is needed to produce energy and energy is necessary to make water available for use. As our population grows, the demands for both resources will only increase. Smart development of energy resources will identify, consider, and minimize potential impacts to water resources.

Natural gas, particularly shale gas, is an abundant U.S. energy resource that will be vital to meeting future energy demand and to enabling the nation to transition to greater reliance on renewable energy sources.

Shale gas development both requires significant amounts of water and is conducted in proximity to valuable surface and ground water. Hence, it is important to reconcile the concurrent and related demands for local and regional water resources, whether for drinking water, wildlife habitat, recreation, agriculture, industrial or other uses.

Because shale gas development in the United States is occurring in areas that have not previously experienced oil and gas production, the GWPC has recognized a need for credible, factual information on shale gas resources, technologies for developing these resources, the regulatory framework under which development takes place, and the practices used to mitigate potential impacts on the environment and nearby communities. also addresses nonwater issues that may be of interest to citizens, government officials, water supply and use professionals, and other interested parties.

Each state has laws and regulations to ensure the wise use of its natural resources and to protect the environment. The GWPC has conducted a separate study to summarize state oil and gas program requirements that are designed to protect water resources. These two studies complement one other and together provide a body of information that can serve as a basis for fact-based dialogue on how shale gas development can proceed in an environmentally responsible manner under the auspices of state regulatory programs.

This Shale Gas Primer was intended to be an accurate depiction of current factors and does not represent the view of any individual state. Knowledge about shale gas development will continue to evolve. The GWPC welcomes insights that readers may have about the Primer and the relationship of shale gas development to water resources.

Scott Kell, President,
Ground Water Protection Council

EXECUTIVE SUMMARY

Natural gas production from hydrocarbon rich shale formations, known as "shale gas," is one of the most rapidly expanding trends in onshore domestic oil and gas exploration and production today. In some areas, this has included bringing drilling and production to regions of the country that have seen little or no activity in the past. New oil and gas developments bring change to the environmental and socio-economic landscape, particularly in those areas where gas development is a new activity. With these changes have come questions about the nature of shale gas development, the potential environmental impacts, and the ability of the current regulatory structure to deal with this development. Regulators, policy makers, and the public need an objective source of information on which to base answers to these questions and decisions about how to manage the challenges that may accompany shale gas development.

Natural gas plays a key role in meeting U.S. energy demands. Natural gas, coal and oil supply about 85% of the nation's energy, with natural gas supplying about 22% of the total. The percent contribution of natural gas to the U.S. energy supply is expected to remain fairly constant for the next 20 years.

The United States has abundant natural gas resources. The Energy Information Administration estimates that the U.S. has more than 1,744 trillion cubic feet (tcf) of technically recoverable natural gas, including 211 tcf of proved reserves (the discovered, economically recoverable fraction of the original gas-in-place). Technically recoverable unconventional gas (shale gas, tight sands, and coalbed methane) accounts for 60% of the onshore recoverable resource. At the U.S. production rates for 2007, about 19.3 tcf, the current recoverable resource estimate provides enough natural gas to supply the U.S. for the next 90 years. Separate estimates of the shale gas resource extend this supply to 116 years.

Natural gas use is distributed across several sectors of the economy. It is an important energy source for the industrial, commercial and electrical generation sectors, and also serves a vital role in residential heating. Although forecasts vary in their outlook for future demand for natural gas, they all have one thing in common: natural gas will continue to play a significant role in the U.S. energy picture for some time to come.

The lower 48 states have a wide distribution of highly organic shales containing vast resources of natural gas. Already, the fledgling Barnett Shale play in Texas produces 6% of all natural gas produced in the lower 48 States. Three factors have come together in recent years to make shale gas production economically viable: 1) advances in horizontal drilling, 2) advances in hydraulic fracturing, and, perhaps most importantly, 3) rapid increases in natural gas prices in the last several years as a result of significant supply and demand pressures. Analysts have estimated that by 2011 most new reserves growth (50% to 60%, or approximately 3 bcf/day) will come from unconventional shale gas reservoirs. The total recoverable gas resources in four new shale gas plays (the Haynesville, Fayetteville,

Marcellus, and Woodford) may be over 550 tcf. Total annual production volumes of 3 to 4 tcf may be sustainable for decades. This potential for production in the known onshore shale basins, coupled with other unconventional gas plays, is predicted to contribute significantly to the U.S.'s domestic energy outlook.

Shale gas is present across much of the lower 48 States. Exhibit ES-1 shows the approximate locations of current producing gas shales and prospective shales. The most active shales to date are the Barnett Shale, the Haynesville/Bossier Shale, the Antrim Shale, the Fayetteville Shale, the Marcellus Shale, and the New Albany Shale. Each of these gas shale basins is different and each has a unique set of exploration criteria and operational challenges. Because of these differences, the development of shale gas resources in each of these areas faces potentially unique opportunities and challenges.

The development and production of oil and gas in the U.S., including shale gas, are regulated under a complex set of federal, state, and local laws that address every aspect of exploration and operation. All of the laws, regulations, and permits that apply to conventional oil and gas exploration and production activities also apply to shale gas development. The U.S. Environmental Protection Agency administers most of the federal laws, although development on federally-owned land is managed primarily by the Bureau of Land Management (part of the Department of the Interior) and the U.S. Forest Service (part of the Department of Agriculture). In addition, each state in which oil and gas is produced has one or more regulatory agencies that permit wells, including their design, location, spacing, operation, and abandonment, as well as environmental activities and discharges, including water management and disposal, waste management and disposal, air emissions, underground injection, wildlife impacts, surface disturbance, and worker health and safety. Many of the federal laws are implemented by the states under agreements and plans approved by the appropriate federal agencies.

A series of federal laws governs most environmental aspects of shale gas development. For example, the Clean Water Act regulates surface discharges of water associated with shale gas drilling and production, as well as storm water runoff from production sites. The Safe Drinking Water Act regulates the underground injection of fluids from shale gas activities. The Clean Air Act limits air emissions from engines, gas processing equipment, and other sources associated with drilling and production. The National Environmental Policy Act (NEPA) requires that exploration and production on federal lands be thoroughly analyzed for environmental impacts. Most of these federal laws have provisions for granting "primacy" to the states (i.e., state agencies implement the programs with federal oversight).

State agencies not only implement and enforce federal laws; they also have their own sets of state laws to administer. The states have broad powers to regulate, permit, and enforce all shale gas development activities—the drilling and fracture of the well, production operations, management and disposal of wastes, and abandonment and plugging of the well. State regulation of the environmental practices related to shale gas development, usually with federal oversight, can more effectively address the regional and state-specific character of the activities, compared to one-size-fits-all regulation at the federal level. Some of these specific factors include: geology, hydrology, climate, topography, industry characteristics, development history, state legal structures, population density, and local economics. State laws often add additional levels of environmental protection and requirements. Also, several states have their own versions of the federal NEPA law, requiring environmental assessments and reviews at the state level and extending those reviews beyond federal lands to state and private lands.

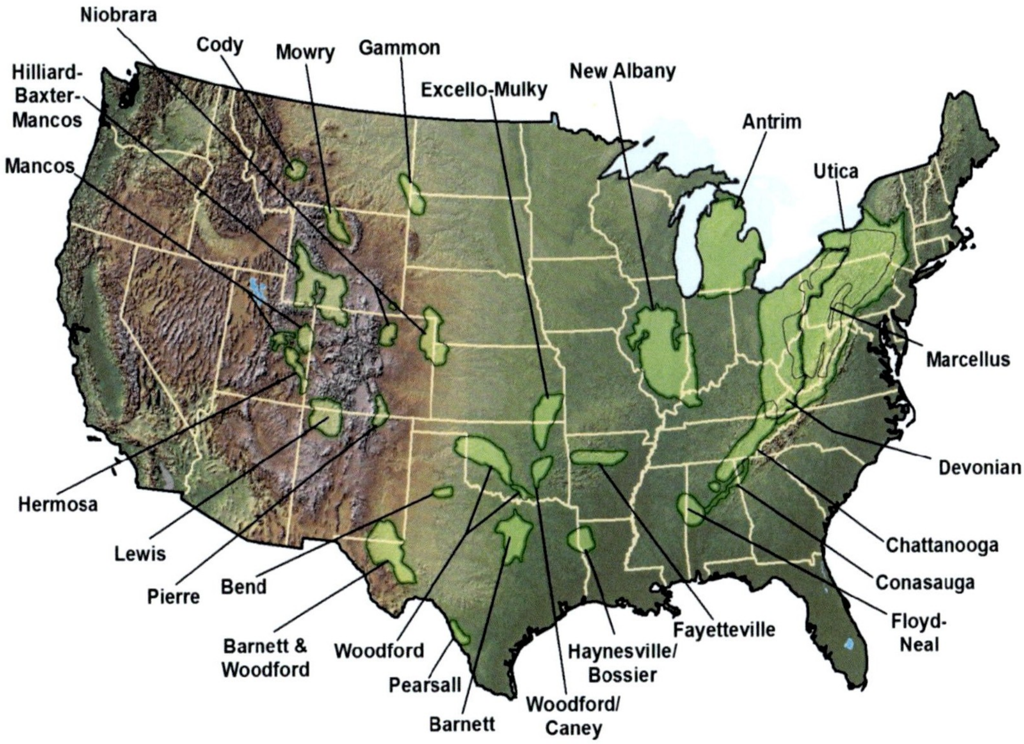

Exhibit ES.1. United States Shale Basins

A key element in the emergence of shale gas production has been the refinement of cost-effective horizontal drilling and hydraulic fracturing technologies. These two processes, along with the implementation of protective environmental management practices, have allowed shale gas development to move into areas that previously would have been inaccessible. Accordingly, it is important to understand the technologies and practices employed by the industry and their ability to prevent or minimize the potential effects of shale gas development on human health and the environment and on the quality of life in the communities in which shale gas production is located.

Modern shale gas development is a technologically driven process for the production of natural gas resources. Currently, the drilling and completion of shale gas wells includes both vertical and horizontal wells. In both kinds of wells, casing and cement are installed to protect fresh and treatable water aquifers. The emerging shale gas basins are expected to follow a trend similar to the Barnett Shale play with increasing numbers of horizontal wells as the plays mature. Shale gas operators are increasingly relying on horizontal well completions to optimize recovery and well economics. Horizontal drilling provides more exposure to a formation than does a vertical well. This increase in reservoir exposure creates a number of advantages over vertical wells drilling. Six to eight horizontal wells drilled from only one well pad can access the same reservoir volume as sixteen vertical wells. Using multi-well pads can also significantly reduce the overall number of well pads, access roads, pipeline routes, and production facilities required, thus minimizing habitat disturbance, impacts to the public, and the overall environmental footprint.

The other technological key to the economic recovery of shale gas is hydraulic fracturing, which involves the pumping of a fracturing fluid under high pressure into a shale formation to generate fractures or cracks in the target rock formation. This allows the natural gas to flow out of the shale to the well in economic quantities. Ground water is protected during the shale gas fracturing process by a combination of the casing and cement that is installed when the well is drilled and the thousands of feet of rock between the fracture zone and any fresh or treatable aquifers. For shale gas development, fracture fluids are primarily water based fluids mixed with additives that help the water to carry sand proppant into the fractures. Water and sand make up over 98% of the fracture fluid, with the rest consisting of various chemical additives that improve the effectiveness of the fracture job. Each hydraulic fracture treatment is a highly controlled process designed to the specific conditions of the target formation.

The amount of water needed to drill and fracture a horizontal shale gas well generally ranges from about 2 million to 4 million gallons, depending on the basin and formation characteristics. While these volumes may seem very large, they are small by comparison to some other uses of water, such as agriculture, electric power generation, and municipalities, and generally represent a small percentage of the total water resource use in each shale gas area. Calculations indicate that water use for shale gas development will range from less than 0.1% to 0.8% of total water use by basin. Because the development of shale gas is new in some areas, these water needs may still challenge supplies and infrastructure. As operators look to develop new shale gas plays, communication with local water planning agencies, state agencies, and regional water basin commissions can help operators and communities to coexist and effectively manage local water resources. One key to the successful development of shale gas is the identification of water supplies capable of meeting the needs of a development company for drilling and fracturing water without interfering with community needs. While a variety of options exist, the conditions of obtaining water are complex and vary by region.

After the drilling and fracturing of the well are completed, water is produced along with the natural gas. Some of this water is returned fracture fluid and some is natural formation water. Regardless of the source, these produced waters that move back through the wellhead with the gas represent a stream that must be managed. States, local governments, and shale gas operators seek to manage produced water in a way that protects surface and ground water resources and, if possible, reduces future demands for fresh water. By pursuing the pollution prevention hierarchy of "Reduce, Re-use, and Recycle" these groups are examining both traditional and innovative approaches to managing shale gas produced water. This water is currently managed through a variety of mechanisms, including underground injection, treatment and discharge, and recycling. New water treatment technologies and new applications of existing technologies are being developed and used to treat shale gas produced water for reuse in a variety of applications. This allows shale gas-associated produced water to be viewed as a potential resource in its own right.

Some soils and geologic formations contain low levels of naturally occurring radioactive material (NORM). When NORM is brought to the surface during shale gas drilling and production operations, it remains in the rock pieces of the drill cuttings, remains in solution with produced water, or, under certain conditions, precipitates out in scales or sludges. The radiation from this NORM is weak and cannot penetrate dense materials such as the steel used in pipes and tanks.

Because the general public does not come into contact with gas field equipment for extended periods, there is very little exposure risk from gas field NORM. To protect gas field workers, OSHA requires employers to evaluate radiation hazards, post caution signs and provide personal protection equipment when radiation doses could exceed regulatory standards. Although regulations vary by state, in general, if NORM concentrations are less than regulatory standards, operators are allowed to dispose of the material by methods approved for standard gas field waste. Conversely, if NORM concentrations are above regulatory limits, the material must be disposed of at a licensed facility. These regulations, standards, and practices ensure that shale gas operations present negligible risk to the general public and to workers with respect to potential NORM exposure.

Although natural gas offers a number of environmental benefits over other sources of energy, particularly other fossil fuels, some air emissions commonly occur during exploration and production activities. Emissions may include NO_x, volatile organic compounds, particulate matter, SO_2, and methane. EPA sets standards, monitors the ambient air across the U.S., and has an active enforcement program to control air emissions from all sources, including the shale gas industry. Gas field emissions are controlled and minimized through a combination of government regulation and voluntary avoidance, minimization, and mitigation strategies.

The primary differences between modern shale gas development and conventional natural gas development are the extensive uses of horizontal drilling and high-volume hydraulic fracturing. The use of horizontal drilling has not introduced any new environmental concerns. In fact, the reduced number of horizontal wells needed coupled with the ability to drill multiple wells from a single pad has significantly reduced surface disturbances and associated impacts to wildlife, dust , noise, and traffic. Where shale gas development has intersected with urban and industrial settings, regulators and industry have developed special practices to alleviate nuisance impacts, impacts to sensitive environmental resources, and interference with existing businesses. Hydraulic fracturing has been a key technology in making shale gas an affordable addition to the Nation's energy supply, and the technology has proved to be an effective stimulation technique. While some challenges exist with water availability and water management, innovative regional solutions are emerging that allow shale gas development to continue while ensuring that the water needs of other users are not affected and that surface and ground water quality is protected. Taken together, state and federal requirements along with the technologies and practices developed by industry serve to reduce environmental impacts from shale gas operations.

INTRODUCTION

Natural gas production from hydrocarbon-rich shale formations, known as "shale gas", is one of the most rapidly expanding trends in onshore domestic oil and gas exploration and production today. In some areas, this has included bringing drilling and production to regions of the country that have seen little or no activity in the past. New oil and gas developments bring changes to the environmental and socio-economic landscape, particularly in those areas where gas development is a new activity. With these changes have come questions about the nature of shale gas development, the potential environmental impacts, and the ability of the current regulatory structure to deal with this development. Regulators, policy makers, and the public need an objective source of information on which to base answers to these questions and decisions about how to manage the challenges that may accompany shale gas development.

This Primer endeavors to provide much of that information. It describes the importance of shale gas in meeting the future energy needs of the United States (U.S.), including its role in alternative energy strategies and reducing greenhouse gas (GHG) emissions. The Primer provides an overview of modern shale gas development, as well as a summary of federal, state, and local regulations applicable to the natural gas production industry, and describes environmental considerations related to shale gas development.

The Primer is intended to serve as a technical summary document, including geologic information on the shale gas basins in the U.S. and the methods of shale gas development. By providing an overview of the regulatory framework and the environmental considerations associated with shale gas development, it will also help facilitate the minimization and mitigation of adverse environmental impacts. By so doing, the Primer can serve as an instrument to facilitate informed public discussions and to support sound policy-making decisions by government.

The Importance of Shale Gas

The Role of Natural Gas in the United States' Energy Portfolio

Natural gas plays a key role in meeting U.S. energy demands. Natural gas, coal and oil supply about 85% of the nation's energy, with natural gas supplying about 22% of the total[1] (Exhibit 1[2]). The percent contribution of natural gas to the U.S. energy supply is expected to remain fairly constant for the next 20 years.

The United States has abundant natural gas resources. The Energy Information Administration (EIA) estimates that the U.S. has more than 1,744 trillion cubic feet (tcf) of technically recoverable natural gas, including 211 tcf of proved reserves (the discovered, economically recoverable fraction of the original gas-in-place)[3,4]. Navigant Consulting estimates that technically recoverable unconventional gas (shale gas, tight sands, and coalbed natural gas) accounts for 60% of the onshore recoverable resource[5]. At the U.S. production rates for 2007, about 19.3 tcf, the current recoverable resource estimate provides enough natural gas to supply the U.S. for the next 90 years[6]. Note that historically, estimates of the size of the total recoverable resource have grown over time as knowledge of the resource has improved and recovery technology has advanced. Unconventional gas resources are a prime example of this trend.

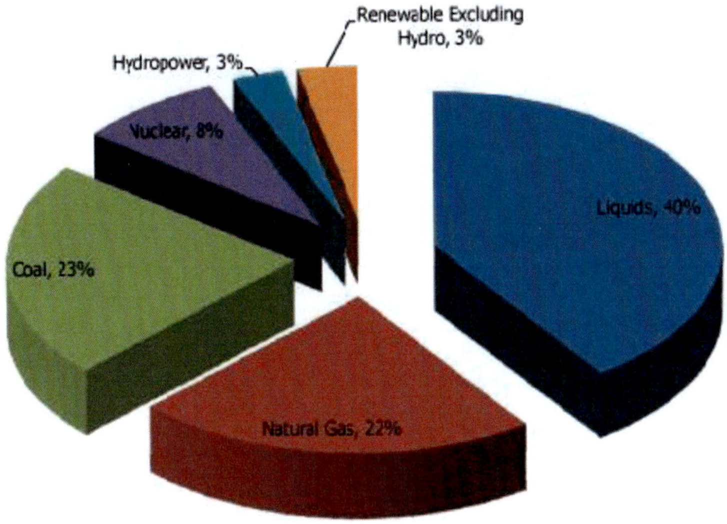

Source: EIA, 2008.

Exhibit 1. United States Energy Consumption by Fuel (2007)

WHAT IS A TCF?

Natural gas is generally priced and sold in units of a thousand cubic feet (Mcf, using the Roman numeral for one thousand). Units of a trillion cubic feet (tcf) are often used to measure large quantities, as in resources or reserves in the ground, or annual national energy consumption. A tcf is one billion Mcf and is enough natural gas to:

- *Heat 15 million homes for one year;*
- *Generate 100 billion kilowatt-hours of electricity;*
- *Fuel 12 million natural gas-fired vehicles for one year.*

Natural gas use is distributed across several sectors of the economy (Exhibit 2[7]). It is an important energy source for the industrial, commercial and electrical generation sectors, and also serves a vital role in residential heating[8]. Although forecasts vary in their outlook for future demand for natural gas, they all have one thing in common: natural gas will continue to play a significant role in the U.S. energy picture for some time to come[9].

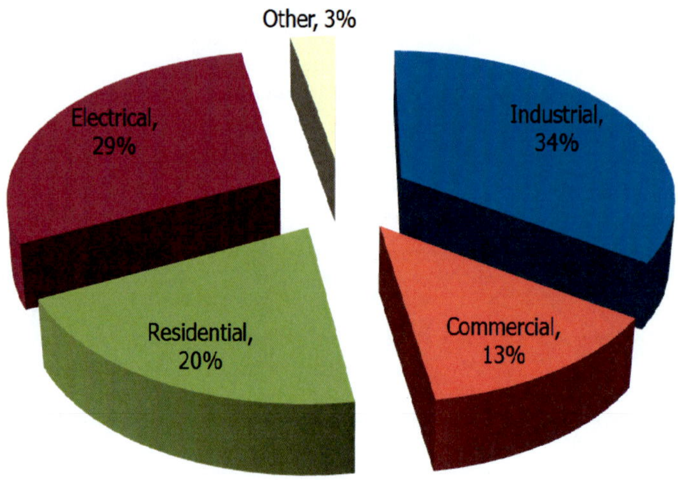

Source: EAI, 2009

Exhibit 2. Natural Gas Use by Sector

Natural gas, due to its clean-burning nature and economical availability, has become a very popular fuel for the generation of electricity[10]. In the 1970s and 80s, the choice for the majority of electric utility generators was primarily coal or nuclear power; but, due to economic, environmental, technological, and regulatory changes, natural gas has become the fuel of choice for many new power plants. In 2007, natural gas was 39.1%[11] of electric industry productive capacity.

Natural gas is also the fuel of choice for a wide range of industries. It is a major fuel source for pulp and paper, metals, chemicals, petroleum refining, and food processing. These five industries alone account for almost three quarters of industrial natural gas use[12] and together employ four million people in the U.S.[13] Natural gas is also a feedstock for a variety of products, including plastics, chemicals, and fertilizers. For many products, there is no economically viable substitute for natural gas. Industrial use of natural gas accounted for 6.63 tcf of demand in 2007 and is expected to grow to 6.82 tcf by 2030.

However, natural gas is being consumed by the U.S. economy at a rate that exceeds domestic production and the gap is increasing[14]. Half of the natural gas consumed today is produced from wells drilled within the last 3.5 years[15]. Despite possessing a large resource endowment, the U.S. consumes natural gas at a rate requiring rapid replacement of reserves. It is estimated that the gap between demand and domestic supply will grow to nearly 9 tcf by the year 2025[16]. However, it is believed by many that unconventional natural gas resources such as shale gas can significantly alter that balance.

Half of the natural gas consumed today is produced from wells drilled within the last 3.5 years.

Exhibit 3[17] shows a comparison of production, consumption, and import trends for natural gas in the U.S. with demand increasingly exceeding conventional domestic production. Without domestic shale gas and other unconventional gas production, the gap between demand and domestic production will widen even more, leaving imports to fill the need. Worldwide consumption of natural gas is also increasing; therefore the U.S. can anticipate facing an increasingly competitive market for these imports.

This increased reliance on foreign sources of energy could pose at least two problems for the U.S.: 1) it would serve to decrease our energy security; and 2) it could create a multi-billion dollar outflow to foreign interests, thus making such funds unavailable for domestic investment.

The Advantages of Natural Gas

In the 1800s and early 1900s, natural gas was mainly used to light streetlamps and the occasional house. However, with a vastly improved distribution network and advancements in technology, natural gas is now being used in many ways. One reason for the widespread use of natural gas is its versatility as a fuel. Its high British thermal unit (Btu) content and a well-developed infrastructure make it easy to use in a number of applications.

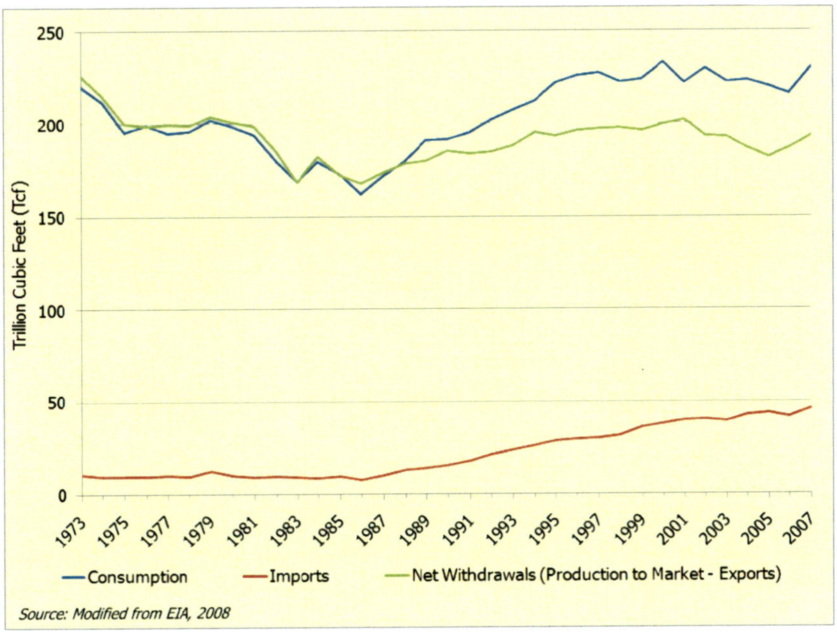

Exhibit 3. Comparison of Production, Consumption and Import Trends for Natural Gas in the United States

Another factor that makes natural gas an attractive energy source is its reliability. Eighty-four percent of the natural gas consumed in the U.S. is produced in the U.S., and ninety-seven percent of the gas used in this country is produced in North America[18]. Thus, the supply of natural gas is not dependent on unstable foreign countries and the delivery system is less subject to interruption.

A key advantage of natural gas is that it is efficient and clean burning[19]. In fact, of all the fossil fuels, natural gas is by far the cleanest burning. It emits approximately half the carbon dioxide (CO_2) of coal along with low levels of other air pollutants[20]. The combustion byproducts of natural gas are mostly CO_2 and water vapor, the same compounds people exhale when breathing. Coal and oil are composed of much more complex organic molecules with greater nitrogen and sulfur content. Their combustion byproducts include larger quantities of CO_2, nitrogen oxides (NO_x), sulfur dioxide (SO_2) and particulate ash (Exhibit 421). By comparison, the combustion of natural gas liberates very small amounts of SO_2 and NOx, virtually no ash, and lower levels of CO_2, carbon monoxide (CO), and other hydrocarbons[22].

Because natural gas emits only half as much CO_2 as coal and approximately 30% less than fuel oil, it is generally considered to be central to energy plans focused on the reduction of GHG emissions[23]. According to the EIA in its report "Emissions of Greenhouse Gases in the United States 2006," 82.3% of GHG emissions in the U.S. in 2006 came from CO_2 as a direct result of fossil fuel combustion[24]. Since CO_2 makes up a large fraction of U.S. GHG emissions, increasing the role of natural gas in U.S. energy supply relative to other fossil fuels would result in lower GHG emissions.

Of all the fossil fuels, natural gas is by far the cleanest burning.

Exhibit 4. Combustion Emissions (Pounds/Billion BTU of Energy Input)

Air Pollutant	Combusted Source		
	Natural Gas	Oil	Coal
Carbon dioxide (CO_2)	117,000	164,000	208,000
Carbon monoxide (CO)	40	33	208
Nitrogen oxides (NO_x)	92	448	457
Sulfur dioxide (SO_2)	0.6	1,122	2,591
Particulates (PM)	7.0	84	2,744
Formaldehyde	0.750	0.220	0.221
Mercury (Hg)	0.000	0.007	0.016

Sources: EIA, 1998

Although there is rapidly increasing momentum to reduce dependence on fossil fuels in the U.S. and elsewhere, the transition to sustainable renewable energy sources will no doubt require considerable time, effort and investment in order for these sources to become economical enough to supply a significant portion of the nation's energy consumption. Indeed, the EIA estimates that fossil fuels (oil, gas, and coal) will supply 82.1% of the nation's energy needs in 2030[25]. Since natural gas is the cleanest burning of the fossil fuels, an environmental benefit could be realized by shifting toward proportionately greater reliance on natural gas until such time as sources of alternative energy are more efficient, economical, and widely available.

Additionally, the march towards sustainable renewable energy sources, such as wind and solar, requires that a supplemental energy source be available when weather conditions and electrical storage capacity prove challenging[26]. Such a backstop energy source must be widely available on near instantaneous demand. The availability of extensive natural gas transmission and distribution pipeline systems makes natural gas uniquely suitable for this role[27]. Thus, natural gas is an integral facet of moving forward with alternative energy options. With the current emphasis on the potential effects of air emissions on global climate change, air quality, and visibility, cleaner fuels like natural gas are an important part of our nation's energy future[28].

Natural Gas Basics

Natural gas is a combination of hydrocarbon gases consisting primarily of methane (CH4), and lesser percentages of butane, ethane, propane, and other gases[29,30]. It is odorless, colorless, and, when ignited, releases a significant amount of energy[31]. Exhibit 5[32] shows the typical compositional range of natural gas produced in the U.S.

Natural gas is found in rock formations (reservoirs) beneath the earth's surface; in some cases it may be associated with oil deposits. Exploration and production companies explore for these deposits by using complex technologies to identify prospective drilling locations. Once extracted, the natural gas is processed to eliminate other gases, water, sand, and other impurities. Some hydrocarbon gases, such as butane and propane, are captured and separately marketed. Once it has been processed, the cleaned natural gas is distributed through a system of pipelines across thousands of miles[33]. It is through these pipelines that natural gas is transported to its endpoint for residential, commercial, and industrial use.

Natural gas is measured in either volumetric or energy units. As a gas, it is measured by the volume it displaces at standard temperatures and pressures, usually expressed in cubic feet. Gas companies generally measure natural gas in thousands of cubic feet (Mcf), millions of cubic feet (MMcf), or billions of cubic feet (bcf), and estimate resources such as original gas-in-place in trillions of cubic feet (tcf).

Calculating and tracking natural gas by volume is useful, but it can also be measured as a source of energy. Similar to other forms of energy, natural gas can be computed and presented in British thermal units (Btu). One Btu is the quantity of heat required to raise the temperature of one pound of water by one degree Fahrenheit at normal pressure[34]. There are about 1,000 Btus in one cubic foot of natural gas delivered to the consumer[35]. Natural gas distribution companies typically measure the gas delivered to a residence in 'therms' for billing purposes[36]. A therm is equal to 100,000 Btus—approximately 100 cubic feet—of natural gas[37].

Unconventional Gas

The U.S. increased its natural gas reserves by 6% from 1970 to 2006, producing approximately 725 tcf of gas during that period[38]. This increase is primarily a result of advancements in technology, resulting in an increase in economically recoverable reserves (reserves becoming proven) that were previously thought to be uneconomic[39].

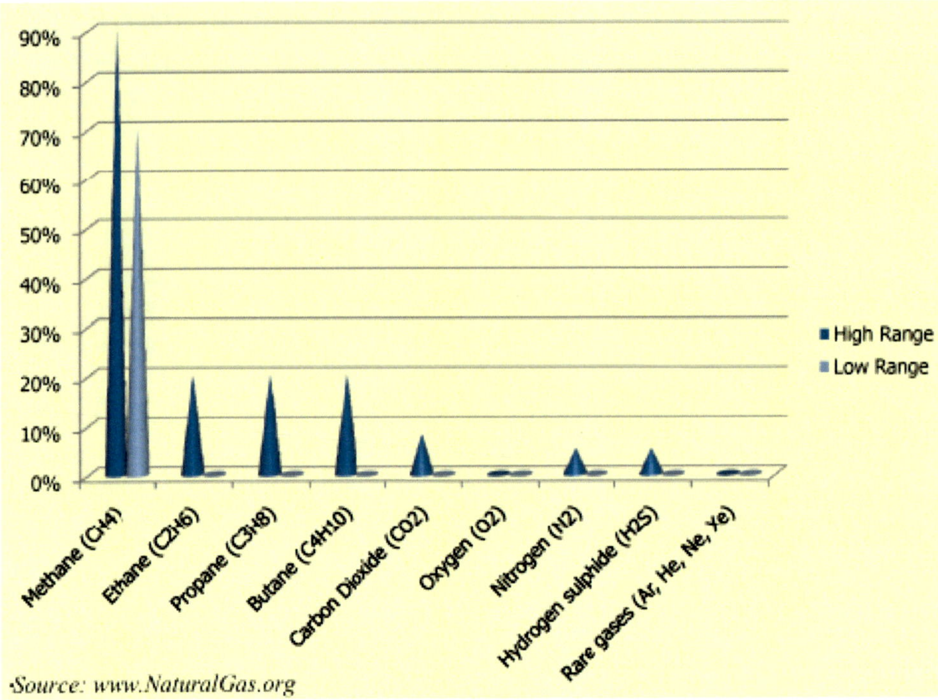

Exhibit 5. Typical Composition of Natural Gas

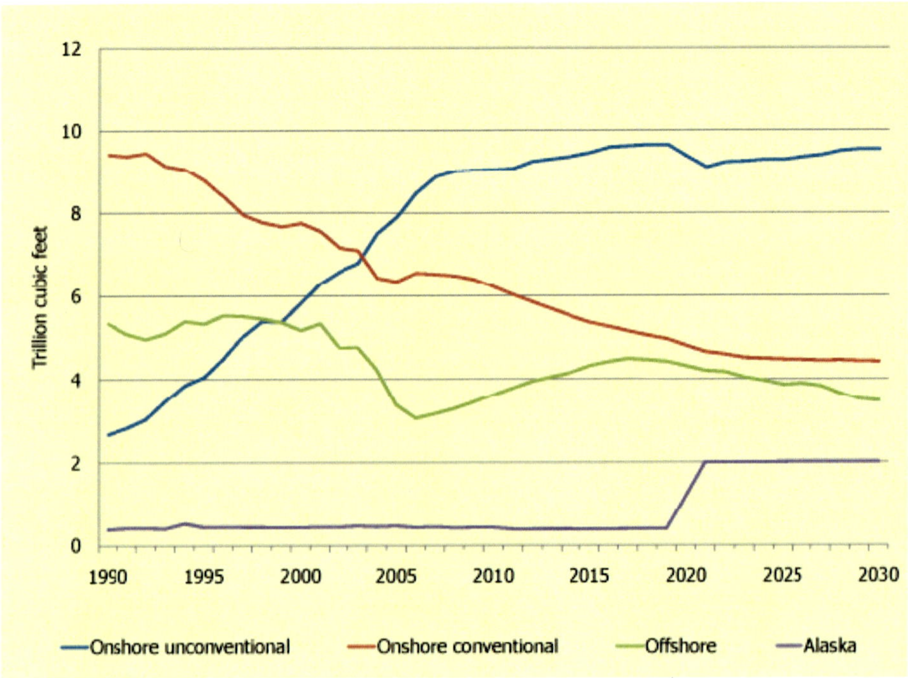

Source: EIA, 2008

Exhibit 6. Natural Gas Production by Source (Tcf/Year)

In 2007, Texas, Wyoming, and Colorado were the states with the greatest additions to proved gas reserves for the year; these additions were from shale gas, tight sands, and coalbed methane, all of which are unconventional gas plays[40]. Similarly, the states of Texas (30%) and Wyoming (12%) had the greatest volume of proved gas reserves in the U.S. in 2007—again, both primarily as a result of developing unconventional natural gas plays[41].

Overall, unconventional natural gas is anticipated to become an ever-increasing portion of the U.S. proved reserves, while conventional gas reserves are declining[42]. Over the last decade, production from unconventional sources has increased almost 65%, from 5.4 trillion cubic feet per year (tcf/yr) in 1998 to 8.9 tcf/yr in 2007 (Exhibit 6). This means unconventional production now accounts for 46% of the total U.S. production[43].

Unconventional production now accounts for 46% of the total U.S. production.

The Role of Shale Gas in Unconventional Gas

The lower 48 states have a wide distribution of highly organic shales containing vast resources of natural gas (Exhibit 7[44]). Already, the fledgling Barnett Shale play in Texas produces 6% of all natural gas produced in the lower 48 states[45]. Improved drilling and fracturing technologies have contributed considerably to the economic potential of shale gas. This potential for production in the known onshore shale basins, coupled with other unconventional gas plays, is predicted to contribute significantly to the U.S.'s domestic energy outlook. Exhibit 8[46] shows the projected contribution of shale gas to the overall unconventional gas production in the U.S. in bcf/day.

Three factors have come together in recent years to make shale gas production economically viable: 1) advances in horizontal drilling, 2) advances in hydraulic fracturing, and, perhaps most importantly, 3) rapid increases in natural gas prices in the last several years as a result of significant supply and demand pressures. Advances in the pre-existing technologies of directional drilling and hydraulic fracturing set the stage for today's horizontal drilling and fracturing techniques, without which many of the unconventional natural gas plays would not be economical. As recently as the late 1990s, only 40 drilling rigs (6% of total active rigs in the U.S.) in the U.S. were capable of onshore horizontal drilling; that number grew to 519 rigs (28% of total active rigs in the U.S.) by May 2008[47].

It has been suggested that the rapid growth of unconventional natural gas plays has not been captured by recent resource estimates compiled by the EIA and that, therefore, their resource estimates do not accurately reflect the contribution of shale gas[48]. Since 1998, annual production has consistently exceeded the EIA's forecasts of unconventional gas production. A great deal of this increase is attributable to shale gas production, particularly from the Barnett Shale in Texas. The potential for most other shale gas plays in the U.S. is just emerging. Taking this into consideration, Navigant, adding their own analysis of shale gas resources to other national resource estimates, has estimated that U.S. total natural gas resources (proved plus unproved technically recoverable) are 1,680 tcf to 2,247 tcf, or 87 to 116 years of production at 2007 U.S. production levels. This compares with EIA's national resource estimate of 1,744 tcf, which is within the Navigant range. Navigant has estimated that shale gas comprises 28% or more of total estimated technically recoverable gas resources in the

U.S.[49]. Exhibit 950 depicts the daily production (in MMcf/day) from each of the currently active shale gas plays.

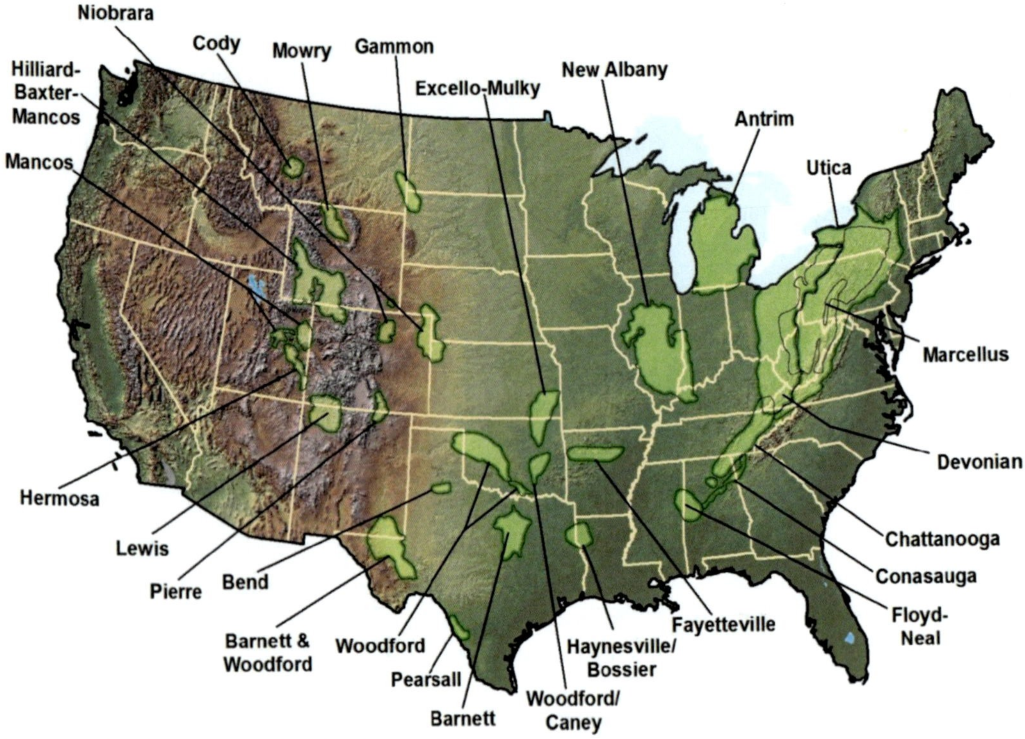

Source: ALL Consulting, Modified from USGS & other sources

Exhibit 7. United States Shale Gas Basins

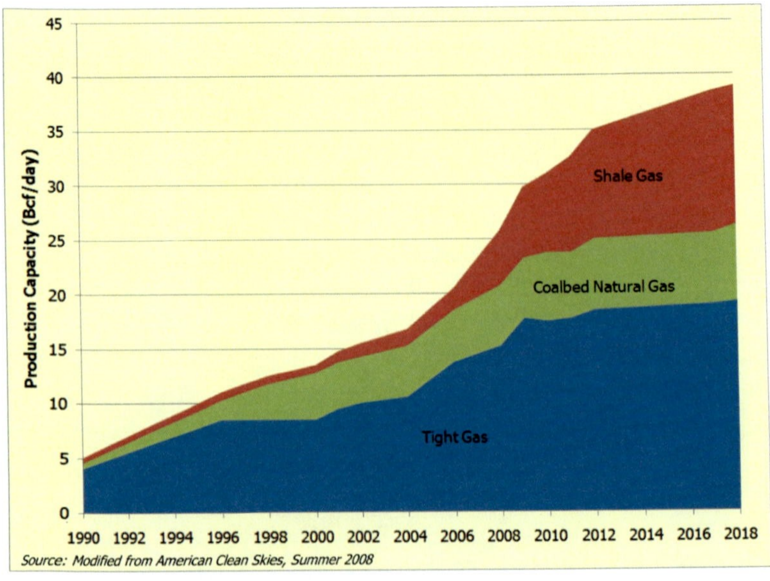

Exhibit 8. United States Unconventional Gas Outlook (Bcf/Day)

Three factors have come together in recent years to make shale gas production economically viable: 1) advances in horizontal drilling, 2) advances in hydraulic fracturing, and, perhaps most importantly, 3) rapid increases in natural gas prices.

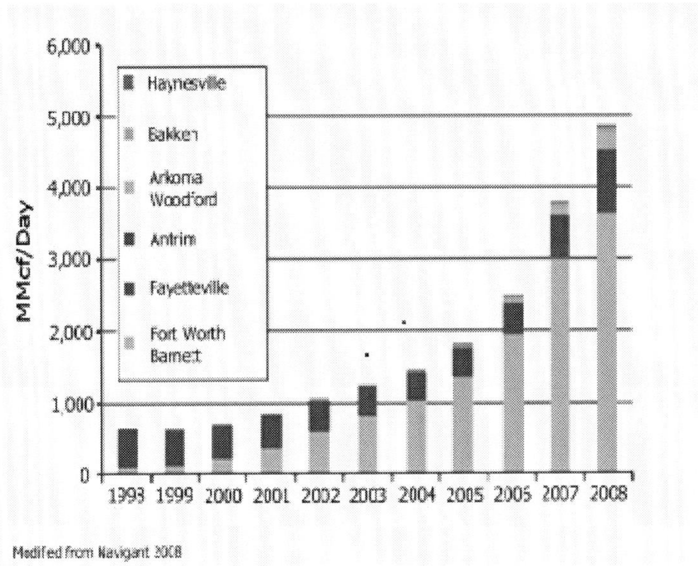

Exhibit 9. Trends in Shale Gas Production (MMcf/Day)

As with most resource estimates, especially emerging resources such as unconventional natural gas, these estimates are likely to change over time. In addition, there are a variety of organizations making resource and future production estimates for shale gas. These analyses use different assumptions, data, and methodologies. Therefore, one may come across a wide range of numbers for projected shale gas recovery, both nationally and by basin. These shale gas resource estimates are likely to change as new information, additional experience, and advances in technology become available.

Shale gas resource estimates are likely to change as new information, additional experience, and advances in technology become available.

Analysts have estimated that by 2011 most new reserves growth (50% to 60%, or approximately 3 bcf/day) will come from unconventional shale gas reservoirs[51]. The total recoverable gas resources from 4 emerging shale gas plays (the Haynesville, Fayetteville, Marcellus, and Woodford) may be over 550 tcf[52]. Total annual production volumes of 3 to 4 tcf may be sustainable for decades. An additional benefit of shale gas plays is that many exist in areas previously developed for natural gas production and, therefore, much of the necessary pipeline infrastructure is already in place. Many of these areas are also proximal to the nation's population centers thus potentially facilitating transportation to consumers. However, additional pipelines will have to be built to access development in areas that have not seen gas production before[53].

Looking Forward

Considering natural gas's clean-burning nature, the nation's domestic natural gas resources, and the presence of supporting infrastructure, the development of domestic shale gas reserves will be an important component of the U.S.'s energy portfolio for many years. Recent successes in a variety of geologic basins have created the opportunity for shale gas to be a strategic part of the nation's energy and economic growth[54].

The Environmental Considerations section of this Primer describes how improvements in horizontal drilling and hydraulic fracturing technologies have opened the door to the economic recovery of shale gas. It also discusses additional practices that have allowed development of areas that might previously have been inaccessible due to environmental constraints or restrictions on disturbances in both urban and rural settings. By using horizontal drilling, operators have been able to reduce the extent of surface impact commonly associated with multiple vertical wells drilled from multiple well pads; equivalent well coverage can be achieved through drilling fewer horizontal wells from a single well pad. This can result in a significant reduction in surface disturbances: fewer well pads, fewer roads, reduced traffic, fewer pipelines, and fewer surface facilities. In urban settings, this can mean less impact on nearby populations and businesses. In rural settings, this can mean fewer consequences for wildlife habitats, agricultural resources, and surface water bodies.

Recent successes and improvements in a variety of geologic basins have created the opportunity for shale gas to be a strategic part of the nation's energy and economic growth.

Other practices that are now commonly used for drilling, particularly in urban settings, include: the use of sound walls and blankets to reduce noise, the use of directional or shielded lighting to reduce nighttime disturbance to nearby residences and businesses, the use of pipelines to transport water resulting in reduced truck traffic, and the use of solar-powered telemetry devices to monitor gas production resulting in reduced personnel visits to well sites. Such practices are used in specific locations or situations that call for them, and are not appropriate everywhere, but where needed, they provide opportunities for safe, environmentally sound development that may not have been possible without them.

These technologies and practices, along with the increasing gas prices of the last few years, have provided the means by which shale gas can be economically recovered. Improvements in reducing the overall footprint and level of disturbance from drilling and completion activities have provided the industry with the methods for moving forward with development in new areas that were previously inaccessible.

Shale Gas Development in the United States

Shale formations across the U.S. have been developed to produce natural gas in small but continuous volumes since the earliest years of gas development. The first producing gas well in the U.S. was completed in 1821 in Devonian-aged shale near the town of Fredonia, New York[55]. The natural gas from this first well was used by town residents for lighting[56]. Early supplies of natural gas were derived from shallow gas wells that were not complicated to drill and from natural gas seeps[57]. The shallow wells and seeps were capable of producing small amounts of natural gas that were used for illuminating city streets and households[58]. These early gas wells played a key part in bringing illumination to the cities and towns of the eastern U.S.[59].

The first producing gas well in the U.S. was completed in 1821 in Devonian-aged shale near the town of Fredonia, New York.

Other shale gas wells followed the Fredonia well with the first field-scale development of shale gas from the Ohio Shale in the Big Sandy Field of Kentucky during the 1920s[60]. The Big Sandy Field has recently experienced a renewed growth and currently is a 3,000-square-mile play encompassing five counties[61]. By the 1930s, gas from the Antrim Shale in Michigan had experienced moderate development; however, it was not until the 1980s that development began to expand rapidly to the point that it has now reached nearly 9,000 wells[62]. It was also during the 1980s that one of the nation's most active natural gas plays initially kicked off in the area around Fort Worth, Texas[63]. The play was the Barnett Shale, and its success grabbed the industry's attention. Large-scale hydraulic fracturing, a process first developed in Texas in the 1950s, was first used in the Barnett in 1986; likewise, the first Barnett horizontal well was drilled in 1992[64]. Through continued improvements in the techniques and technology of hydraulic fracturing, development of the Barnett Shale has accelerated[65]. In the ensuing two decades, the science of shale gas extraction has matured into a sophisticated process that utilizes horizontal drilling and sequenced, multi-stage hydraulic fracturing technologies. As the Barnett Shale play has matured, natural gas producers have been looking to extrapolate the lessons learned in the Barnett to the other shale gas formations present across the U.S. and Canada[66].

In addition to the Barnett Play, a second shale play with greater oil production has also been advancing techniques related to horizontal wells and hydraulic fracturing. The Bakken Shale of the Williston Basin of Montana and North Dakota has seen a similar growth rate to the Barnett. The Bakken is another technical play in which the development of this unconventional resource has benefitted from the technological advances in horizontal wells and hydraulic fracturing[67]. In April 2008, the United States Geological Survey (USGS) released an updated assessment of the undiscovered technically recoverable reserves for this shale play estimating there are 3.65 billion barrels (bbls) of oil, 1.85 tcf of associated natural gas, and 148 million bbls of natural gas liquids in the play[68].

The combination of sequenced hydraulic fracture treatments and horizontal well completions has been crucial in facilitating the expansion of shale gas development. Prior to the successful application of these two technologies in the Barnett Shale, shale gas resources in many basins had been overlooked because production was not viewed as economically feasible[69]. The low natural permeability of shale has been the limiting factor to the production of shale gas resources because it only allows minor volumes of gas to flow naturally to a wellbore[70]. The characteristic of low-matrix permeability represents a key difference between shale and other gas reservoirs. For gas shales to be economically produced, these restrictions must be overcome[71]. The combination of reduced economics and low permeability of gas shale formations historically caused operators to bypass these formations and focus on other resources[72].

Shale Gas – Geology

Shale gas is natural gas produced from shale formations that typically function as both the reservoir and source for the natural gas. In terms of its chemical makeup, shale gas is typically a dry gas primarily composed of methane (90% or more methane), but some formations do produce wet gas. The Antrim and New Albany formations have typically produced water and gas[73]. Gas shales are organic-rich shale formations that were previously regarded only as source rocks and seals for gas accumulating in the stratigraphically-associated sandstone and carbonate reservoirs of traditional onshore gas development[74]. Shale is a sedimentary rock that is predominantly comprised of consolidated clay-sized particles. Shales are deposited as mud in low-energy depositional environments such as tidal flats and deep water basins where the fine-grained clay particles fall out of suspension in these quiet waters. During the deposition of these very fine-grained sediments, there can also be deposition of organic matter in the form of algae-, plant-, and animal-derived organic debris[75]. The naturally tabular clay grains tend to lie flat as the sediments accumulate and subsequently become compacted as a result of additional sediment deposition. This results in mud with thin laminar bedding that lithifies (solidifies) into thinly layered shale rock. The very fine sheet-like clay mineral grains and laminated layers of sediment result in a rock that has limited horizontal permeability and extremely limited vertical permeability. Typical unfractured shales have matrix permeabilities on the order of 0.01 to 0.00001 millidarcies[76]. This low permeability means that gas trapped in shale cannot move easily within the rock except over geologic expanses of time (millions of years).

The natural layering and fracturing of shales can be seen in outcrop. Exhibit 10 shows a typical shale outcrop which reveals the natural bedding planes, or layers, of the shale and near-vertical natural fractures that can cut across the naturally horizontal bedding planes. Although the vertical fractures shown in this picture are naturally occurring, artificial fractures induced by hydraulic fracture stimulation in the deep subsurface reservoir rock would have a similar appearance.

The low permeability of shale causes it to be classified as an unconventional reservoir for gas (or in some cases, oil) production. These low permeability, often organic-rich units are also thought to be the source beds for much of the hydrocarbons produced in these basins[77]. Gas reservoirs are classified as conventional or unconventional for the following reasons:

1. **Conventional reservoirs** – Wells in conventional gas reservoirs produce from sands and carbonates (limestones and dolomites) that contain the gas in interconnected pore spaces that allow flow to the wellbore. Much like a kitchen sponge, the gas in the pores can move from one pore to another through smaller pore-throats that create permeable flow through the reservoir. In conventional natural gas reservoirs, the gas is often sourced from organic-rich shales proximal to the more porous and permeable sandstone or carbonate.

Source: ALL Consulting, 2008

Exhibit 10. Marcellus Shale Outcrop

2. **Unconventional reservoirs** – Wells in unconventional reservoirs produce from low permeability (tight) formations such as tight sands and carbonates, coal, and shale. In unconventional gas reservoirs, the gas is often sourced from the reservoir rock itself (tight gas sandstone and carbonates are an exception). Because of the low permeability of these formations, it is typically necessary to stimulate the reservoir to create additional permeability. Hydraulic fracturing of a reservoir is the preferred stimulation method for gas shales. Differences between the three basic types of unconventional reservoirs include:

    1. **Tight Gas** – Wells produce from regional low-porosity sandstones and carbonate reservoirs. The natural gas is sourced (formed) outside the reservoir and migrates into the reservoir over time (millions of years)[78]. Many of these wells are drilled horizontally and most are hydraulically fractured to enhance production.

    2. **Coal Bed Natural Gas (CBNG)** – Wells produce from the coal seams which act as source and reservoir of the natural gas[79]. Wells frequently produce water as well as natural gas. Natural gas can be sourced by thermogenic alterations of coal or by biogenic action of indigenous microbes on the coal. There are some horizontally drilled CBNG wells and some that receive hydraulic fracturing treatments. However, some CBNG reservoirs are also underground sources of drinking water and as such there are restrictions on hydraulic fracturing. CBNG wells are mostly shallow as the coal matrix does not have the strength to maintain porosity under the pressure of significant overburden thickness.

3. **Shale Gas** – Wells produce from low permeability shale formations that are also the source for the natural gas. The natural gas volumes can be stored in a local macro-porosity system (fracture porosity) within the shale, or within the micro-pores of the shale[80], or it can be adsorbed onto minerals or organic matter within the shale[81]. Wells may be drilled either vertically or horizontally and most are hydraulically fractured to stimulate production. Shale gas wells can be similar to other conventional and unconventional wells in terms of depth, production rate, and drilling.

Sources of Natural Gas

Shale gas is both created and stored within the shale bed. Natural gas (methane) is generated from the organic matter that is deposited with and present in the shale matrix.

In order for a shale to have economic quantities of gas it must be a capable source rock. The potential of a shale formation to contain economic quantities of gas can be evaluated by identifying specific source rock characteristics such as total organic carbon (TOC), thermal maturity, and kerogen analysis. Together, these factors can be used to predict the likelihood of the prospective shale to produce economically viable volumes of natural gas. A number of wells may need to be analyzed in order to sufficiently characterize the potential of a shale formation, particularly if the geologic basin is large and there are variations in the target shale zone.

Shale Gas in the United States

Shale gas is present across much of the lower 48 States. Exhibit 7 shows the approximate locations of current producing gas shales and prospective shales. The most active shales to date are the Barnett Shale, the Haynesville/Bossier Shale, the Antrim Shale, the Fayetteville Shale, the Marcellus Shale, and the New Albany Shale. The following discussion provides a summary of basic information regarding these shale gas plays.

Each of these gas shale basins is different and each has a unique set of exploration criteria and operational challenges. Because of these differences, the development of shale gas resources in each of these areas faces potentially unique challenges. For example, the Antrim and New Albany Shales are shallower shales that produce significant volumes of formation water unlike most of the other gas shales. Development of the Fayetteville Shale is occurring in rural areas of north central Arkansas, while development of the Barnett Shale is focused in the area of Forth Worth, Texas, in an urban and suburban environment.

KEY GAS RESOURCE TERMS

Proved Reserves: *That portion of recoverable resources that is demonstrated by actual production or conclusive formation tests to be technically, economically, and legally producible under existing economic and operating conditions.*

Technically Recoverable Resources: *The total amount of resource, discovered and undiscovered, that is thought to be recoverable with available technology, regardless of economics.*

Original Gas-In-Place: *The entire volume of gas contained in the reservoir, regardless of the ability to produce it.*

As new technologies are developed and refined, shale gas plays once believed to have limited economic viability are now being re-evaluated. Exhibit 11 summarizes the key characteristics of the most active shale gas plays across the U.S. This exhibit supplies data related to the character of the shale and also provides a means to compare some of the key characteristics that are used to evaluate the different gas shale basins. Note that estimates of the shale gas resource, especially the portion that is technically recoverable, are likely to increase over time as new data become available from additional drilling, as experience is gained in producing shale gas, as understanding of the resource characteristics increases, and as recovery technologies improve.

The Barnett Shale

The Barnett Shale is located in the Fort Worth Basin of north-central Texas. It is a Mississippian-age shale occurring at a depth of 6,500 feet to 8,500 feet (Exhibit 11 and Exhibit 13[131]) and is bounded by limestone formations above (Marble Falls Limestone) and below (Chappel Limestone) (Exhibit 12).

With over 10,000 wells drilled to date, the Barnett Shale is the most prominent shale gas play in the U.S.[132]. It has been a showcase for modern tight-reservoir development typical of gas shales in the U.S.[133]. The development of the Barnett Shale has been a proving ground for combining the technologies of horizontal drilling and large-volume hydraulic fracture treatments. Drilling operations continue expanding the play boundaries outward; at the same time, operations have turned towards infill drilling to increase the amount of gas recovered[134]. Horizontal well completions in the Barnett are occurring at well spacing ranging from 60 to 160 acres per well (Exhibit 11).

The Barnett Shale covers an area of about 5,000 square miles with an approximate thickness ranging from 100 feet (ft) to more than 600 ft (Exhibit 11). The original gas-in-place estimate for the Barnett Shale is 327 tcf with estimated technically recoverable resources of 44 tcf (Exhibit 11). The gas content is the highest among the major shale plays, ranging from 300 standard cubic feet per ton (scf/ton) to 350 scf/ton of rock (Exhibit 11).

The Fayetteville Shale

The Fayetteville Shale is situated in the Arkoma Basin of northern Arkansas and eastern Oklahoma over a depth range of 1,000 ft to 7,000 ft (Exhibit 15[135] and Exhibit 11). The Fayetteville Shale is a Mississippian-age shale bounded by limestone (Pitkin Limestone) above and sandstone (Batesville Sandstone) below (Exhibit 14).

Exhibit 11. Comparison of Data for the Gas Shales in the United States

Gas Shale Basin	Barnett	Fayetteville	Haynesville	Marcellus	Woodford	Antrim	New Albany
Estimated Basin Area, square miles	5,000	9,000	9,000	95,000	11,000	12,000	43,500
Depth, ft	6, 500 - 8,500[82]	1,000 - 7,000[83]	10,500 - 13,500[84]	4,000 - 8,500[85]	6,000 - 11,000[86]	600 - 2,200[87]	500 - 2,000[88]
Net Thickness, ft	100 - 600[89]	20 -200[90]	200[91] -300[92]	50 - 200[93]	120 - 220[94]	70 - 120[95]	50 - 100[96]
Depth to Base of Treatable Water[#], ft	~1200	~500[97]	~400	~850	~400	~300	~400
Rock Column Thickness between Top of Pay and Bottom of Treatable Water, ft	5,300 - 7,300	500 - 6,500	10,100 - 13,100	2,125 - 7650	5,600 - 10,600	300 - 1,900	100 - 1,600
Total Organic Carbon, %	4.5[98]	4.0 -9.8[99]	0.5 -4.0[100]	3 -12[101]	1 -14[102]	1 -20[103]	1 -25[104]
Total Porosity, %	4 -5[105]	2 -8[106]	8 -9[107]	10[108]	3 -9[109]	9[110]	10 -14[111]
Gas Content, scf/ton	300 - 350[112]	60 - 220[113]	100 - 330[114]	60 - 100[115]	200 - 300[116]	40 - 100[117]	40 -80[118]
Water Production, Barrels water/day	N/A	N/A	N/A	N/A	N/A	5 -500[119]	5 -500[120]
Well spacing, acres	60 - 160[121]	80 -160	40 -560[122]	40 - 160[123]	640[124]	40 - 160[125]	80[126]
Original Gas-In-Place, tcf[127]	327	52	717	1,500	23	76	160
Technically Recoverable Resources, tcf[128]	44	41.6	251	262	11.4	20	19.2

NOTE: Information presented in this table, such as Original Gas-in-Place and Technically Recoverable Resources, is presented for general comparative purposes only. The numbers provided are based on the sources shown and this research did not include a resource evaluation. Rather, publically available data was obtained from a variety of sources and is presented for general characterization and comparison. Resource estimates for any basin may vary greatly depending on individual company experience, data available at the time the estimate was performed, and other factors. Furthermore, these estimates are likely to change as production methods and technologies improve.

Mcf = thousands of cubic feet of gas

scf = standard cubic feet of gas

tcf = trillions of cubic feet of gas

= For the Depth to base of treatable water data, the data was based on depth data from state oil and gas agencies and state geological survey data.

N/A = Data not available

Exhibit 12. Stratigraphy of the Barnett Shale

	Period	Group/Unit
Permian	Leonardian	Clear Fork Grp
		Wichita Grp
	Wolfcampian	Cisco Grp
Pennsylvanian	Virgilian	
	Missourian	Canyon Grp
	Desmoinesian	Strawn Grp
	Atokan	Bend Grp Marble Falls Limestone
	Morrowan	
	Chesterian - Meramecian	Barnett Shale
Mississippian	Osagean	Chappel Limestone
		Viola Limestone
Ordovician	Canadian	Simpson Grp Ellenburger Grp

Source: Hayden and Pursell, 2005[129] AAPG, 1987 [130]

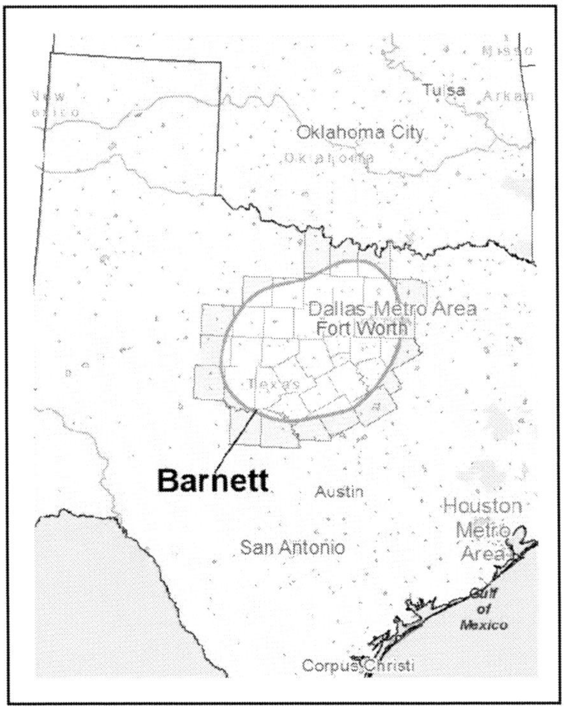

Source: ALL Consulting, 2009

Exhibit 13. Barnett Shale in the Fort Worth Basin

Development of the Fayetteville began in the early 2000s as gas companies that had experienced success in the Barnett Shale of the Fort Worth Basin identified parallels between it and the Mississippian-aged Fayetteville Shale in terms of age and geologic character[136]. Lessons learned from the horizontal drilling and hydraulic fracturing techniques employed in the Barnett, when adapted to development of the Fayetteville Shale, made this play economical[137]. Between 2004 and 2007 the number of gas wells drilled annually in the Fayetteville shale jumped from 13 to more than 600, and gas production for the shale increased from just over 100 MMcf/yr to approximately 88.85 bcf/yr[138]. With over 1,000 wells in production to date, the Fayetteville Shale is currently on its way to becoming one of the most active plays in the U.S.[139].

Exhibit 14. Stratigraphy of the Fayetteville Shale

<table>
<tr><th></th><th>Period</th><th colspan="2">Group/Unit</th></tr>
<tr><td rowspan="12">CARBONIFEROUS</td><td rowspan="5">Pennsylvanian</td><td colspan="2">Atoka</td></tr>
<tr><td colspan="2">Bloyd</td></tr>
<tr><td rowspan="2">Hale</td><td>Prairie</td></tr>
<tr><td>Grove</td></tr>
<tr><td colspan="2">Cane Hill</td></tr>
<tr><td rowspan="6">Mississippian</td><td colspan="2">(IMO)</td></tr>
<tr><td colspan="2">Pitkin</td></tr>
<tr><td colspan="2">Fayetteville</td></tr>
<tr><td colspan="2">Batesville</td></tr>
<tr><td colspan="2">Moorefield</td></tr>
<tr><td colspan="2">Boone</td></tr>
</table>

Source: Hillwood, 2007[140]

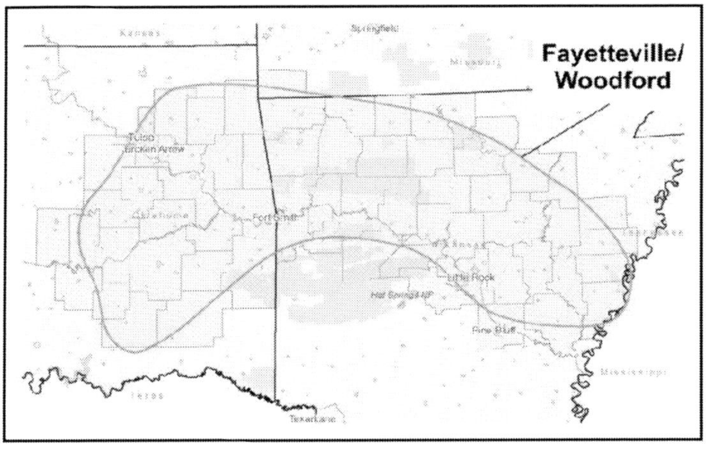

Source: ALL Consulting, 2009

Exhibit 15. Fayetteville Shale in the Arkoma Basin

The area of the Fayetteville Shale play is nearly double that of the Barnett Shale at 9,000 square miles, with well spacing ranging from 80 to 160 acres per well, and pay zone thickness averaging between 20 ft and 200 ft (Exhibit 11). The gas content for the Fayetteville Shale has been measured at 60 to 220 scf/ton, which is less than the 300 to 350 scf/ton gas content of the Barnett. The lower gas content of the Fayetteville, as compared to the Barnett, results in lower estimates of the original gas-in-place and technically recoverable resources: 52 tcf and 41.6 tcf respectively (Exhibit 11).

The Haynesville Shale

The Haynesville Shale (also known as the Haynesville/Bossier) is situated in the North Louisiana Salt Basin in northern Louisiana and eastern Texas with depths ranging from 10,500 ft to 13,500 ft (Exhibit 17[141] and Exhibit 11). The Haynesville is an Upper Jurassic-age shale bounded by sandstone (Cotton Valley Group) above and limestone (Smackover Formation) below (Exhibit 16).

Exhibit 16. Stratigraphy of the Haynesville Shale

<table>
<tr><th colspan="2">Period</th><th>Group/Unit</th></tr>
<tr><td rowspan="11">Cretaceous</td><td rowspan="17">Upper</td><td>Navarro</td></tr>
<tr><td>Taylor</td></tr>
<tr><td>Austin</td></tr>
<tr><td>Eagle Ford</td></tr>
<tr><td>Tuscaloosa</td></tr>
<tr><td></td></tr>
<tr><td>Washita</td></tr>
<tr><td>Fredericksburg</td></tr>
<tr><td>Trinity Group</td></tr>
<tr><td></td></tr>
<tr><td>Nuevo Leon</td></tr>
<tr><td rowspan="8">Jurassic</td><td></td></tr>
<tr><td>Cotton Valley Group</td></tr>
<tr><td></td></tr>
<tr><td>Haynesville</td></tr>
<tr><td>Smackover</td></tr>
<tr><td>Norphlet</td></tr>
<tr><td>Middle</td><td>Louann</td></tr>
<tr><td>Lower</td><td>Werner</td></tr>
<tr><td>Triassic</td><td>Upper</td><td>Eagle Mills</td></tr>
</table>

Source: Johnson et al, 2000[143]

In 2007, after several years of drilling and testing, the Haynesville Shale made headlines as a potentially significant gas reserve, although the full extent of the play will only be known after several more years of development are completed[142].

The Haynesville Shale covers an area of approximately 9,000 square miles with an average thickness of 200 ft to 300 ft (Exhibit 11). The thickness and areal extent of the Haynesville has allowed operators to evaluate a wider variety of spacing intervals ranging from 40 to 560 acres per well (Exhibit 11). Gas content estimates for the play are 100 scf/ton to 330 scf/ton. The Haynesville formation has the potential to become a significant shale gas resource for the U.S. with original gas-in-place estimates of 717 tcf and technically recoverable resources estimated at 251 tcf (Exhibit 11).

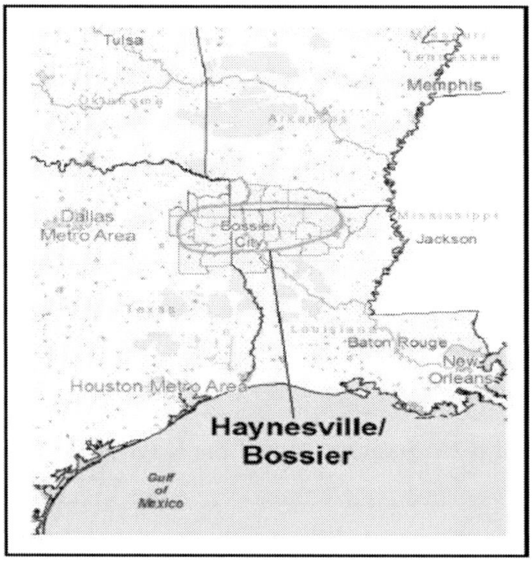

Source: ALL Consulting, 2009

Exhibit 17. Haynesville SHALE in the Texas & Louisiana Basin

The Marcellus Shale

The Marcellus Shale is the most expansive shale gas play, spanning six states in the northeastern U.S. (Exhibit 19[144]). The estimated depth of production for the Marcellus is between 4,000 ft and 8,500 ft (Exhibit 11). The Marcellus Shale is a Middle Devonian-age shale bounded by shale (Hamilton Group) above and limestone (Tristates Group) below (Exhibit 18).

Following an increase in gas prices, triggered by the Natural Gas Policy Act (NGPA) of 1978, Devonian shale gas development rose in the early- to mid-1980s in the northeast, but decreasing gas prices resulted in uneconomical wells and declining production through the 1990s[145]. In 2003, Range Resources Corporation drilled the first economically producing wells into the Marcellus formation in Pennsylvania using horizontal drilling and hydraulic fracturing techniques similar to those used in the Barnett Shale formation of Texas[146]. Range Resources began producing this formation in 2005. As of September 2008, there were a total of 518 wells permitted in Pennsylvania in the Marcellus shale and 277 of the approved wells had been drilled[147].

The Marcellus Shale covers an area of 95,000 square miles at an average thickness of 50 ft to 200 ft (Exhibit 11). While the Marcellus is lower in relative gas content at 60 scf/ton to 100 scf/ton, the much larger area of this play compared to the other shale gas plays results in a higher original gas-in-place estimate of up to 1,500 tcf (Exhibit 11).

Exhibit 18. Stratigraphy of the Marcellus Shale

	Period	Group/Unit	
Penn		Pottsville	
Miss Devonian	**Upper**	Pocono	
		Conewango	
		Conneaut	
		Canadaway	
		West Falls	
		Sonyea	
		Genesee	
	Middle	Tully	
		Hamilton Group	Moscow
			Ludlowville
			Skaneateles
			Marcellus
		Onandaga	
	Lower	Tristates	
		Helderberg	

Source: Arthur et al, 2008[148]

At an average well spacing in the Marcellus is 40 to 160 acres per well (Exhibit 11). The data in Exhibit 11 show technically recoverable resources for the formation to be 262 tcf, although much like the Haynesville, the play's potential estimates are frequently being revised upward due to its early stage of development.

The Woodford Shale

Located in south-central Oklahoma, the Woodford Shale ranges in depth from 6,000 ft to 11,000 ft (Exhibit 21[149] and Exhibit 11). This formation is a Devonian-age shale bounded by limestone (Osage Lime) above and undifferentiated strata below (Exhibit 20).

Recent natural gas production in the Woodford Shale began in 2003 and 2004 with vertical well completions only[150]. However, horizontal drilling has been adopted in the Woodford, as in other shale gas plays, due to its success in the Barnett Shale[151].

The Woodford Shale play encompasses an area of nearly 11,000 square miles (Exhibit 11). The Woodford play is in an early stage of development and is occurring at a spacing interval of 640 acres per well (Exhibit 11). The average thickness of the Woodford Shale varies from 120 ft to 220 ft across the play (Exhibit 11).

Gas content in the Woodford Shale is higher on average than some of the other shale gas plays at 200 scf/ton to 300 scf/ton (Exhibit 11). The original gas-in-place estimate for the Woodford Shale is similar to the Fayetteville Shale at 23 tcf while the technically recoverable resources are 11.4 tcf (Exhibit 11).

The Antrim Shale

The Antrim Shale is located in the upper portion of the lower peninsula of Michigan within the Michigan Basin (Exhibit 23[154]). This Late Devonian-age shale is bounded by shale (Bedford Shale) above and by limestone (Squaw Bay Limestone) below and occurs at depths of 600 ft to 2,200 ft which is more typical of CBNG formations than most gas shales (Exhibit 22 and Exhibit 11).

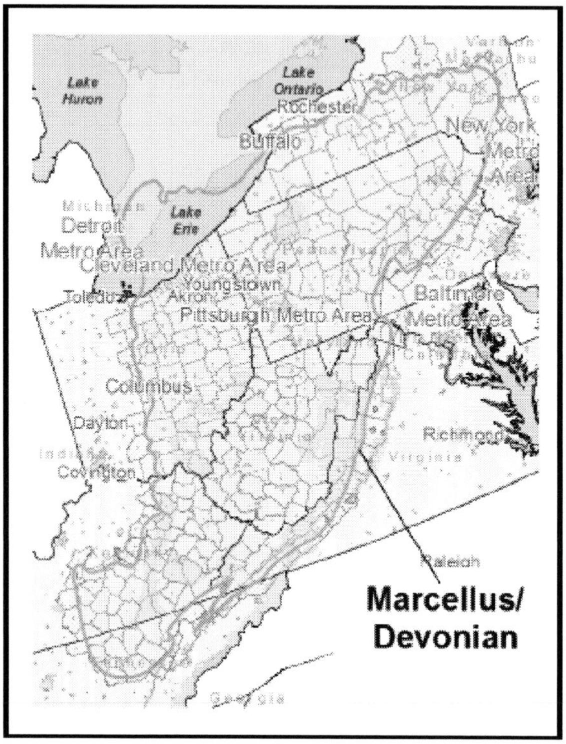

Source: ALL Consulting, 2009

Exhibit 19. Marcellus Shale in the Appalachian Basin

Exhibit 20. Stratigraphy of the Woodford Shale

<table>
<tr><th colspan="2">Period</th><th colspan="2">Group/Unit</th></tr>
<tr><td rowspan="8">Permian</td><td>Ochoan</td><td colspan="2" rowspan="2">Cloyd Chief Fm</td></tr>
<tr><td rowspan="3">Guadalupian</td></tr>
<tr><td colspan="2">White Horse Grp</td></tr>
<tr><td colspan="2">El Reno Grp</td></tr>
<tr><td>Leonardian</td><td colspan="2">Enid Grp</td></tr>
<tr><td rowspan="3">Wolfcampian</td><td colspan="2">Chase Grp</td></tr>
<tr><td colspan="2">Council Grove Grp</td></tr>
<tr><td colspan="2">Admire Grp</td></tr>
<tr><td rowspan="2">Penn.</td><td>Atokan</td><td colspan="2">Atoka Grp</td></tr>
<tr><td>Morrowan</td><td colspan="2">Morrow Grp</td></tr>
<tr><td rowspan="4">Mississippian</td><td>Chesterian</td><td colspan="2">Chester Grp</td></tr>
<tr><td>Meramecian</td><td rowspan="3">Miss Lime</td><td rowspan="2">Meramec Lime</td></tr>
<tr><td>Osagean</td></tr>
<tr><td>Kinderhookian</td><td>Osage Lime</td></tr>
<tr><td rowspan="3">Devonian</td><td>Upper</td><td colspan="2">Woodford Shale</td></tr>
<tr><td>Middle</td><td colspan="2">Undifferentiated</td></tr>
<tr><td>Lower</td><td>Hunton Grp</td><td>Haragan Fm Henryhouse Fm</td></tr>
</table>

Source: Cardott, 2007[152]
AAPG, 1983[153]

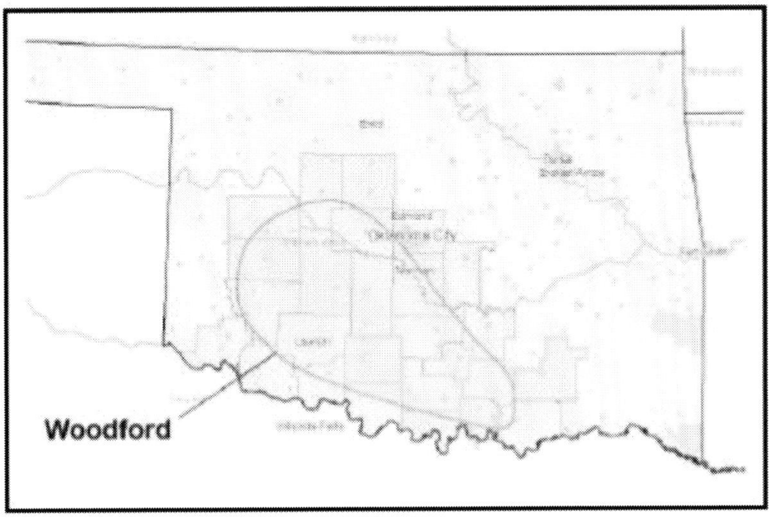

Source: ALL Consulting, 2009

Exhibit 21. Woodford Shale in the Anadarko Basin

Exhibit 22. Stratigraphy of the Antrim Shale

<table>
<tr><th colspan="2">Period</th><th colspan="2">Formation</th></tr>
<tr><td>Quaternary</td><td>Pleistocene</td><td colspan="2">Glacial Drift</td></tr>
<tr><td>Jurassic</td><td>Middle</td><td colspan="2">Ionia Formation</td></tr>
<tr><td rowspan="2">Pennsylvanian</td><td>Late</td><td colspan="2">Grand River Formation</td></tr>
<tr><td>Early</td><td colspan="2">Saginaw Formation</td></tr>
<tr><td rowspan="6">Mississippian</td><td rowspan="3">Late</td><td colspan="2">Parma Formation</td></tr>
<tr><td colspan="2">Bayport Limestone</td></tr>
<tr><td colspan="2">Michigan Formation</td></tr>
<tr><td rowspan="3">Early</td><td colspan="2">Marshall Sandstone</td></tr>
<tr><td colspan="2">Coldwater Shale</td></tr>
<tr><td colspan="2">Sunbury Shale</td></tr>
<tr><td rowspan="7">Devonian</td><td rowspan="7">Late</td><td rowspan="2">Ellsworth Shale</td><td>Berea Sandstone</td></tr>
<tr><td>Bedford Shale</td></tr>
<tr><td>Upper Member</td><td rowspan="4">Antrim Shale</td></tr>
<tr><td>Lachine Member</td></tr>
<tr><td>Paxton Member</td></tr>
<tr><td>Norwood Member</td></tr>
<tr><td colspan="2">Squaw Bay Limestone</td></tr>
</table>

Source: Catacosinos et al, 2000[157]

Aside from the Barnett, the Antrim Shale has been one of the most actively developed shale gas plays with its major expansion taking place in the late 1980s[155].

The Antrim Shale encompasses an area of approximately 12,000 square miles and is characterized by distinct differences from other gas shales: shallow depth, small stratigraphic thickness with average net pay of 70 ft to 120 ft, and greater volumes of produced water in the range of 5 to 500 bbls/day/well[156] (Exhibit 11).

The gas content of the Antrim Shale ranges between 40 scf/ton and 100 scf/ton (Exhibit 11). The original gas-in-place for the Antrim is estimated at 76 tcf with technically recoverable resources estimated at 20 tcf (Exhibit 11). Well spacing ranges from 40 acres to 160 acres per well.

Source: ALL Consulting, 2009

Exhibit 23. Antrim Shale in the Michigan Basin

The New Albany Shale

The New Albany Shale is located in the Illinois Basin in portions of southeastern Illinois, southwestern Indiana, and northwestern Kentucky[159] (Exhibit 25[160]). Similar to the Antrim Shale, the New Albany occurs at depths between 500 ft and 2,000 ft (Exhibit 11) and is a shallower, water-filled shale with a more CBNG-like character than the other gas shales discussed in this section. The New Albany formation is a Devonian- to Mississippian-age shale bounded by limestone above (Rockford Limestone) and below (North Vernon Limestone) (Exhibit 24).

Exhibit 24. Stratigraphy of the New Albany Shale

<table>
<tr><th colspan="2">Period</th><th colspan="2">Formation</th></tr>
<tr><td rowspan="10">Pennsylvanian</td><td></td><td colspan="2">Mattoon</td></tr>
<tr><td rowspan="2">Missourian</td><td colspan="2">Bond</td></tr>
<tr><td colspan="2">Patoka</td></tr>
<tr><td rowspan="5">Desmoinesian</td><td colspan="2">Shelburn</td></tr>
<tr><td colspan="2">Dugger</td></tr>
<tr><td colspan="2">Petersburg</td></tr>
<tr><td colspan="2">Linton</td></tr>
<tr><td colspan="2">Staunton</td></tr>
<tr><td>Atokan</td><td colspan="2">Brazil</td></tr>
<tr><td>Morrowan</td><td colspan="2">Mansfield</td></tr>
<tr><td rowspan="4">Mississippian</td><td rowspan="4">Chesterian</td><td colspan="2">Tobinsport</td></tr>
<tr><td colspan="2">Branchville</td></tr>
<tr><td colspan="2">Tar Springs</td></tr>
<tr><td colspan="2">Glen Dean Limestone</td></tr>
<tr><td rowspan="21"></td><td rowspan="11"></td><td colspan="2">Hardinsburg</td></tr>
<tr><td colspan="2">Haney Limestone</td></tr>
<tr><td colspan="2">Big Clifty</td></tr>
<tr><td colspan="2">Beech Creek Limestone</td></tr>
<tr><td>Cypress</td><td>Elwren</td></tr>
<tr><td colspan="2">Reelsville Limestone</td></tr>
<tr><td colspan="2">Sample</td></tr>
<tr><td colspan="2">Beaver Bend Limestone</td></tr>
<tr><td colspan="2">Bethel</td></tr>
<tr><td colspan="2">Paoli Limestone</td></tr>
<tr><td colspan="2">Ste. Genevieve Ls.</td></tr>
<tr><td rowspan="6">Valmeyeran</td><td colspan="2">St. Louis Limestone</td></tr>
<tr><td colspan="2">Salem Limestone</td></tr>
<tr><td colspan="2">Harrodsburg Limestone</td></tr>
<tr><td>Muldraugh</td><td>Ramp Creek</td></tr>
<tr><td colspan="2">Edwardsville</td></tr>
<tr><td colspan="2">Spickert Knob</td></tr>
<tr><td rowspan="4">Kinderhookian</td><td colspan="2">New Providence Sh.</td></tr>
<tr><td colspan="2"></td></tr>
<tr><td rowspan="2">Rockford Ls</td><td>Coldwater Sh.</td></tr>
<tr><td>Sunbury Sh.</td></tr>
<tr><td rowspan="4">Devonian</td><td rowspan="2">Senecan Chautauquan</td><td rowspan="2">New Albany Shale</td><td>Ellsworth Sh.</td></tr>
<tr><td>Antrim Sh.</td></tr>
<tr><td rowspan="2">Erian</td><td>North Vernon Ls.</td><td>Transverse</td></tr>
<tr><td>Jeffersonville Ls.</td><td>Detroit River</td></tr>
</table>

Source: Indiana Geological Survey, 1986[158]

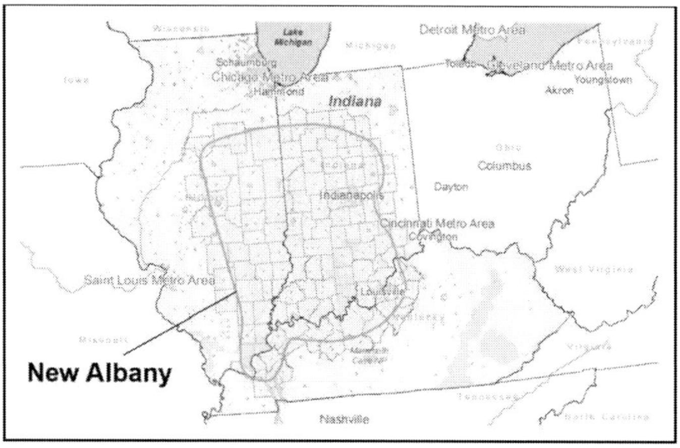

Source: ALL Consulting, 2009

Exhibit 25. New Albany Shale in the Illinois Basin

The New Albany Shale is one of the largest shale gas plays, encompassing an area of approximately 43,500 square miles with approximately 80-acre spacing between wells (Exhibit 11). Similar to the Antrim Shale, the New Albany play has a thinner average net pay thickness of 50 ft to 100 ft and has wells which average 5 to 500 bbls of water per day[161] (Exhibit 11). The measured gas content of the New Albany Shale ranges from 40 scf/ton to 80 scf/ton. The original gas-in-place for the New Albany formation is estimated at 160 tcf with technically recoverable resources estimated at less than 20 tcf (Exhibit 11).

Regulatory Framework

The development and production of oil and gas in the U.S., including shale gas, are regulated under a complex set of federal, state, and local laws that address every aspect of exploration and operation. All of the laws, regulations, and permits that apply to conventional oil and gas exploration and production activities also apply to shale gas development. The U.S. Environmental Protection Agency (EPA) administers most of the federal laws, although development on federally owned land is managed primarily by the Bureau of Land Management (BLM), which is part of the Department of the Interior, and the U.S. Forest Service, which is part of the Department of Agriculture. In addition, each state in which oil and gas is produced has one or more regulatory agencies that permit wells, including their design, location, spacing, operation, and abandonment, as well as environmental activities and discharges, including water management and disposal, waste management and disposal, air emissions, underground injection, wildlife impacts, surface disturbance, and worker health and safety. Many of the federal laws are implemented by the states under agreements and plans approved by the appropriate federal agencies. Those laws and their delegation are discussed below.

Federal Environmental Laws Governing Shale Gas Development

A series of federal laws governs most environmental aspects of shale gas development. For example, the Clean Water Act (CWA) regulates surface discharges of water associated with shale gas drilling and production, as well as storm water runoff from production sites. The Safe Drinking Water Act (SDWA) regulates the underground injection of fluids from shale gas activities. The Clean Air Act (CAA) limits air emissions from engines, gas processing equipment, and other sources associated with drilling and production. The National Environmental Policy Act (NEPA) requires that exploration and production on federal lands be thoroughly analyzed for environmental impacts.

However, federal agencies do not have the resources to administer all of these environmental programs for all the oil and gas sites around the country. Also, as explained below, one set of nation-wide regulations may not always be the most effective way of assuring the desired level of environmental protection. Therefore, most of these federal laws have provisions for granting "primacy" to the states (i.e., state agencies implement the programs with federal oversight). By statute, states may adopt their own standards; however, these must be at least as protective as the federal standards they replace, and may even be more protective in order to address local conditions. Once these state programs are approved by the relevant federal agency (usually the EPA), the state then has primacy jurisdiction.

By statute, states may adopt their own standards; however, these must be at least as protective as the federal standards they replace, and may even be more protective in order to address local conditions.

State Regulation

State regulation of the environmental practices related to shale gas development, usually with federal oversight, can more effectively address the regional and state-specific character of the activities, compared to one-size-fits-all regulation at the federal level[162]. Some of these specific factors include: geology, hydrology, climate, topography, industry characteristics, development history, state legal structures, population density, and local economics. The state agencies that permit these practices and monitor and enforce their laws and regulations may be located in the state Department of Natural Resources (such as in Ohio) or in the Department of Environmental Protection (such as in Pennsylvania). The Texas Railroad Commission regulates oil and gas activity in the nation's largest oil and gas producing state, home to the Barnett Shale. The names and organizational structures vary, but the functions are very similar. Often, multiple agencies are involved, having jurisdiction over different activities and aspects of development.

These state agencies do not only implement and enforce federal laws; they also have their own sets of state laws to administer. These state laws often add additional levels of environmental protection and requirements. Also, several states have their own versions of the federal NEPA law, requiring environmental assessments and reviews at the state level and extending those reviews beyond federal lands to state and private lands.

The states have broad powers to regulate, permit, and enforce all activities—the drilling and fracture of the well, production operations, management and disposal of wastes, and abandonment and plugging of the well.

States have many tools at their disposal to assure that shale gas operations do not adversely impact the environment. The regulation of shale gas drilling and production is a cradle-to-grave approach. The states have broad powers to regulate, permit, and enforce all activities—the drilling and fracture of the well, production operations, management and disposal of wastes, and abandonment and plugging of the well. Different states take different approaches to this regulation and enforcement, but state laws generally give the state oil and gas director or agency the discretion to require whatever is necessary to protect human health and the environmenta[a]. In addition to the general protection regulations, most states have a general prohibition against pollution from oil and gas drilling and production[b]. Most of the state requirements are written into rules or regulations, while some are added to permits on a case-by-case basis as a result of environmental review, on-the-ground inspections, public comments, or commission hearings.

All states require a permit before an operator can drill and operate a gas well. The application for this permit includes all the information about a well's location, construction, operation and reclamation. Agency staff reviews the application for compliance with regulations and to assure adequate environmental safeguards. If necessary, a site inspection will be made before permit approval. Also, most states require operators to post a bond or other financial security when getting a drilling permit to ensure compliance with state regulations and to make sure that there are funds to properly plug the well once production ceases. Another safeguard is that producers generally must notify the state agencies of any significant new activity through a "sundry notice" or a new permit application so that the agency is aware of that activity and can review it[c].

States have implemented voluntary review processes to help ensure that the state programs are as effective as possible. The Ground Water Protection Council (GWPC) has a program to review state implementation of the Underground Injection Control (UIC) program. In addition to the GWPC UIC review, state oil and gas environmental programs other than UIC programs can also be periodically reviewed against a set of guidelines developed by an independent body of state, industry, and environmental stakeholders, known as STRONGER (State Review of Oil and Natural Gas Environmental Regulation, Inc.)[163]. Periodic evaluations of state exploration and production waste management programs have proven useful in improving the effectiveness of those programs and increasing cooperation between federal and state regulatory agencies. To date, 18 states have been reviewed under

[a] An example of this type of provision is the following from Pennsylvania's statute: "[T]he department shall have the authority to issue such orders as are necessary to aid in the enforcement of the provisions of [the oil and gas] act." (58 P.S. section 601.503.).

[b] An example of such language can be found in New York's rules, which state: "The drilling, casing and completion program adopted for any well shall be such as to prevent pollution. Pollution of the land and/or of surface or ground fresh water resulting from exploration or drilling is prohibited." (6 NYCRR Part 554). Another example is the requirement in the rules of the Texas Railroad Commission: "No person conducting activities subject to regulation by the commission may cause or allow pollution of surface or subsurface water in the state." (TAC 16.1.3.8).

[c] See, for example, Louisiana Statewide Order 29-B, section 105, or Texas Administrative Code 16.1.3.5.

the state review guidelines, and several have been reviewed more than once. The STRONGER program has documented the effectiveness of and improvements in these state oil and gas environmental programs[164,165]. The Interstate Oil and Gas Compact Commission (IOGCC) also completed state reviews using earlier versions of the guidelines prior to the formation of STRONGER.

The organization of regulatory agencies within the various oil and gas producing states varies considerably. Some states have several agencies that may oversee some facet of oil and gas operations, especially environmental requirements. These agencies may be in various departments or divisions within the states' organizations. These various approaches have developed over time within each state, and each state tries to create a structure that best serves its citizenry and all of the industries that it must oversee. The one constant is that each oil and gas producing state has one agency with primary responsibility for permitting wells and overseeing general operations. While this agency may work with other agencies in the regulatory process, they can serve as a good source of information about the various agencies that may have jurisdiction over oil and gas activities. Exhibit 26 provides a list of the agencies with primary responsibility for oil and gas regulation in each of the states that have or are likely to have shale gas production.

Local Regulation

In addition to state and federal requirements, additional requirements regarding oil and gas operations may be imposed by other levels of government in specific locations. Entities such as cities, counties, tribes, and regional water authorities may each set operational requirements that affect the location and operation of wells or require permits and approvals in addition to those at the federal or state level.

When operations occur in or near populated areas, local governments may establish ordinances to protect the environment and the general welfare of its citizens. These local ordinances frequently require additional permits for issues such as well placement in flood zones, noise level, set backs from residences or other protected sites, site house-keeping, and traffic. For example, ordinances may set limits on noise levels that may be generated during both daytime and nighttime operations[166,167,168,169].

In some cases, regional water-permitting authorities that have jurisdiction in multiple states have also been established. These federally established authorities have been created to protect the water quality of the entire river basin and to govern uses of the water[170]. Additional approvals and permits may be required for operations in these river basins. For example, the Delaware River Basin Commission (DRBC) covers parts of New York, Pennsylvania, New Jersey and Delaware[171]. Natural gas operators wishing to withdraw water for consumptive use in this basin must first receive a permit from the DRBC, which has the legal authority to fine violators of their rules and regulations.

Exhibit 26. Oil and Gas Regulatory Agencies in Shale Gas States

State	Agency	Web Address
Alabama	Geological Survey of Alabama, State Oil and Gas Board	http://www.ogb.state.al.us/ogb/ogb.html
Arkansas	Arkansas Oil and Gas Commission	http://www.aogc.state.ar.us/
Colorado	Colorado Department of Natural Resources, Oil and Gas Conservation Commission	http://cogcc.state.co.us/
Illinois	Illinois Department of Natural Resources, Division of Oil and Gas	http://dnr.state.il.us/mines/dog/index.htm
Indiana	Indiana Department of Natural Resources, Division of Oil and Gas	http://www.in.gov/dnr/dnroil/
Kentucky	Kentucky Department for Energy Development and Independence, Division of Oil and Gas Conservation	http://www.dogc.ky.gov/
Louisiana	Louisiana Department of Natural Resources, Office of Conservation	http://dnr.louisiana.gov/cons/conserv.ssi
Michigan	Michigan Department of Environmental Quality, Office of Geological Survey	http://www.michigan.gov/deq/0,1607,7-135-3306_28607,00.html
Mississippi	Mississippi State Oil and Gas Board	http://www.ogb.state.ms.us/
Montana	Montana Department of Natural Resources and Conservation, Board of Oil and Gas	http://bogc.dnrc.mt.gov/default.asp
New Mexico	New Mexico Energy, Minerals and Natural Resources Depart-ment, Oil Conservation Division	http://www.emnrd.state.nm.us/OCD/
New York	New York Department of Environmental Conservation, Division of Mineral Resources	http://www.dec.ny.gov/energy/205.html
North Dakota	North Dakota Industrial Commission, Department of Mineral Resources Oil and Gas Division	https://www.dmr.nd.gov/oilgas/
Ohio	Ohio Department of Natural Resources, Division of Mineral Resources Management	http://www.ohiodnr.com/mineral/default/tabid/10352/Default.aspx
Oklahoma	Oklahoma Corporation Com-mission, Oil and Gas Conservation Division	http://www.occ.state.ok.us/Divisions/OG/newweb/og.htm
Pennsylvania	Pennsylvania Department of Environmental Protection, Bu-reau of Oil and Gas Managem-ent	http://www.dep.state.pa.us/dep/DEPUTATE/MINRES/ OILGAS/oilgas.htm
Tennessee	Tennessee Department of Environment and Conserva-tion, State Oil and Gas Board	http://www.tennessee.gov/environment/boards/oilandgas.shtml
Texas	The Railroad Commission of Texas	http://www.rrc.state.tx.us/index.html

The variety of laws governing shale gas exploration and production, and the multitude of federal and state agencies that implement them, can sometimes be confusing. Therefore, the following discussion has been organized according to the various environmental media that are affected by these activities, i.e., water, air, and land. The major laws and programs affecting each of these are discussed below. Additional considerations on federal land and unique state requirements are also covered, along with some of the programs that cut across these environmental media.

Regulation of Impacts on Water Quality

Potential impacts to water quality are primarily regulated under several federal statutes and the accompanying state programs. The primary federal statutes governing water quality issues related to shale gas development are the Clean Water Act, the Safe Drinking Water Act, and the Oil Pollution Act. These statutes and their relationships to shale gas development are discussed below.

Clean Water Act

The Clean Water Act (CWA) is the primary federal law in the U.S. governing pollution of surface water. It was established to protect water quality, and includes regulation of pollutant limits on the discharge of oil- and gas-related produced water. This is conducted through the National Pollutant Discharge Elimination System (NPDES) permitting process. Although EPA sets national standards at the federal level, states and tribal governments can acquire primacy for the NPDES program by meeting EPA's primacy requirements.

The CWA establishes the basic structure for regulating discharges of pollutants into the waters of the U.S. and quality standards for surface waters. The basis of the CWA was enacted in 1948 and was called the Federal Water Pollution Control Act; the Act was significantly reorganized and expanded in 1972. "Clean Water Act" became its common name, with additional amendments made in 1977 and later.

Under the CWA, EPA has implemented pollution control programs such as setting wastewater standards for industry. They have also set water quality standards for a variety of contaminants in surface waters.

The CWA made it unlawful to discharge any pollutant from a point source into the navigable waters of the U.S., unless done in accordance with a specific approved permit. The NPDES permit program controls discharges from point sources that are discrete conveyances, such as pipes or man-made ditches. Industrial, municipal, and other facilities such as shale gas production sites or commercial facilities that handle the disposal or treatment of shale gas produced water must obtain permits if they intend to discharge directly into surface waters[172,173]. Large facilities usually have individual NPDES permits. Discharges from some smaller facilities may be eligible for inclusion under general permits that authorize a category of discharges under the CWA within a geographical area. A general permit is not specifically tailored for an individual discharger. Most oil and gas production facilities with related discharges are authorized under general permits because there are typically numerous sites with common discharges in a geographic area.

A state that meets the federal primacy requirements is allowed to set more stringent state-specific standards for this program. Since individual states can acquire primacy over their respective programs, it is not uncommon to have varying requirements from state to state. This variation can affect how the oil and gas industry manages produced water within a drainage basin located within two or more states, such as the Marcellus shale in the Appalachian Basin. Effluent limitations serve as the primary mechanism under NPDES permits for controlling discharges of pollutants to receiving waters. When developing effluent limitations for an NPDES permit, a permit writer must consider limits based on both the technology available to control the pollutants (i.e., technology-based effluent standards) and the regulations that protect the water quality standards of the receiving water (i.e., water quality-based effluent standards).

The intent of technology-based effluent limits in NPDES permits is to require treatment of effluent concentrations to less than a maximum allowable standard for point source discharges to the specific surface water body. This is based on available treatment technologies, while allowing the discharger to use any available control technique to meet the limits. For industrial (and other non-municipal) facilities, technology-based effluent limits are derived by: 1) using national effluent limitations guidelines and standards established by EPA, or 2) using best professional judgment (BPJ) on a case-by-case basis in the absence of national guidelines and standards.

Prior to the granting of a permit, the authorizing agency must consider the potential impact of every proposed surface water discharge on the quality of the receiving water, not just individual discharges. If the authorizing agency determines that technology-based effluent limits are not sufficient to ensure that water quality standards will be attained in the receiving water, the CWA [Section 303(b)(1)(c)] and NPDES regulations [40 Code of Federal Regulations (CFR) 122.44(d)] require that more stringent limits are imposed as part of the permit[174].

EPA establishes effluent limitation guidelines (ELGs) and standards for different non-municipal (i.e., industrial) categories. These guidelines are developed based on the degree of pollutant reduction attainable by an industrial category through the application of pollution control technologies.

The CWA requires EPA to develop specific effluent guidelines that represent the following:

1. Best conventional technology (BCT) for control of conventional pollutants and applicable to existing dischargers.
2. Best practicable technology (BPT) currently available for control of conventional, toxic and nonconventional pollutants and applicable to existing dischargers.
3. Best available technology (BAT) economically achievable for control of toxic and nonconventional pollutants and applicable to existing dischargers.
4. New source performance standards (NSPS) for conventional pollutants and applicable to new sources.

To date, EPA has established guidelines and standards for more than 50 different industrial categories[175].

The ELGs for Oil and Gas Extraction, which were published in 1979, can be found at 40 CFR Part 435. The onshore subcategory, Subpart C, is applicable to discharges associated with shale gas development and production[176].

The CWA also includes a program to control storm water discharges. The 1987 Water Quality Act (WQA) added Section 402(p) to the CWA requiring EPA to develop and implement a storm water permitting program. EPA developed this program in two phases (Phase I: 1990; Phase II: 1999). Those regulations establish NPDES permit requirements for municipal, industrial, and construction site storm water runoff. The WQA also added Section 402(l)(2) to the CWA specifying that the EPA and states shall not require NPDES permits for uncontaminated storm water discharges from oil and gas exploration, production, processing or treatment operations, or transmission facilities. This exemption applies where the runoff is not contaminated by contact with raw materials or wastes. EPA had previously interpreted the 402(l)(2) exemption as not applying to construction activities of oil and gas development, such as building roads and pads (i.e., an NPDES permit was required)[177].

The Energy Policy Act of 2005 modified the CWA Section 402(l)(2) exemption by defining the excluded oil and gas sector operations to include all oil and gas field activities and operations, including those necessary to prepare a site for drilling and for the movement and placement of drilling equipment. EPA promulgated a rule that implemented this exemption. However, on May 23, 2008, the U.S. Court of Appeals for the Ninth Circuit released a decision vacating the permitting exemption for discharges of sediment from oil and gas construction activities that contribute to violations of the CWA[178]. The court based its decision on the fact that the new rule exempted runoff contaminated with sediment, while the CWA does not exempt such runoff. As a result of the court's decision, storm water discharges contaminated with sediment resulting in a water quality violation require permit coverage under the NPDES storm water permitting program.

While the EPA storm water permitting rule contains a broad exclusion for oil and gas sector construction activities, it is important to note that individual states and Indian tribes may still regulate storm water associated with these activities. EPA has clarified its position that states and tribes may not regulate such storm water discharges under their CWA authority, but are free to regulate under their own independent authorities. EPA states that "[t]his final rule is not intended to interfere with the ability of states, tribes, or local governments to regulate any discharges through a non-NPDES permit program"[179]. In addition to state and tribal regulation, the industry has a voluntary program of Reasonable and Prudent Practices for Stabilization (RAPPS) of oil and gas construction sites[180]. Producers use RAPPS in order to control erosion and sedimentation associated with storm water runoff from areas disturbed by clearing, grading, and excavating activities related to site preparation.

Safe Drinking Water Act

Congress originally passed the Safe Drinking Water Act (SDWA) in 1974 to protect public health by regulating the nation's public drinking water supply. The law was amended in 1986 and 1996 and requires many actions to protect drinking water and its sources, including rivers, lakes, reservoirs, springs, and ground water wells. SDWA authorizes the U.S. EPA to set national health-based standards for drinking water to protect against both naturally occurring and man-made contaminants that may be found in drinking water. EPA, states, and municipal water system agencies then work together to make sure that these standards are met[181].

As one aspect of the protection of drinking water supplies, the SDWA establishes a framework for the Underground Injection Control (UIC) program to prevent the injection of liquid wastes into underground sources of drinking water (USDWs). The EPA and states implement the UIC program, which sets standards for safe waste injection practices and bans certain types of injection altogether. The UIC Program provides these safeguards so that injection wells do not endanger USDWs. The first federal UIC regulations were issued in 1980.

EPA currently groups underground injection wells into five classes for regulatory control purposes, and has a sixth class under consideration. Each class includes wells with similar functions, construction and operating features so that technical requirements can be applied consistently to the class.

1. **Class I** wells may inject hazardous and nonhazardous fluids (industrial and municipal wastes) into isolated formations beneath the lowermost USDW. Because they may inject hazardous waste, Class I wells are the most strictly regulated and are further regulated under the Resource Conservation and Recovery Act (RCRA).
2. **Class II** wells may inject brines and other fluids associated with oil and gas production.
3. **Class III** wells may inject fluids associated with solution mining of minerals.
4. **Class IV** wells may inject hazardous or radioactive wastes into or above a USDW and are banned unless specifically authorized under other statutes for ground water remediation.
5. **Class V** includes all underground injection not included in Classes I-IV. Generally, most Class V wells inject nonhazardous fluids into or above a USDW and are on-site disposal systems, such as floor and sink drains which discharge to dry wells, septic systems, leach fields, and drainage wells. Injection practices or wells that are not covered by the UIC Program include single family septic systems and cesspools as well as non-residential septic systems and cesspools serving fewer than 20 persons that inject ONLY sanitary waste water.
6. **Class VI** has been proposed specifically for the injection of CO_2 for the purpose of sequestration, but has not yet been established.

Most injection wells associated with oil and gas production are Class II wells. These wells may be used to inject water and other fluids (e.g., liquid CO_2) into oil- and gas-bearing zones to enhance recovery, or they may be used to dispose of produced water. The regulation specifically prevents the disposal of waste fluids into USDWs by limiting injection only to formations that are not "underground sources of drinking water." EPA's UIC Program is designed to prevent contamination of water supplies by setting minimum requirements for state UIC Programs. The basic premise of the UIC Program is to prevent contamination of USDWs by keeping injected fluids within the intended injection zone. The injected fluids must not endanger, or have the potential to endanger, a current or future public water supply. The UIC requirements that affect the siting, construction, operation, maintenance, monitoring, testing, and, finally, closure of injection wells have been established to address these concepts. All injection wells require authorization under general rules or specific permits.

The law was written with the understanding that states are best suited to have primary enforcement authority (primacy) for the UIC Program. In the SDWA, Congress cautioned

EPA against a "one-size-fits-all" regulatory scheme, and mandated consideration of local conditions and practices. Section 1421(b)(3)(A) requires that UIC regulations permit or provide consideration of varying geological, hydrological, or historical conditions in different states and in different areas within a state. Section 1425 allows a state to obtain primacy from EPA for oil- and gas-related injection wells, without being required to adopt the complete set of applicable federal UIC regulations. The state must be able to demonstrate that its existing regulatory program is protecting USDWs as effectively as the federal requirements[182].

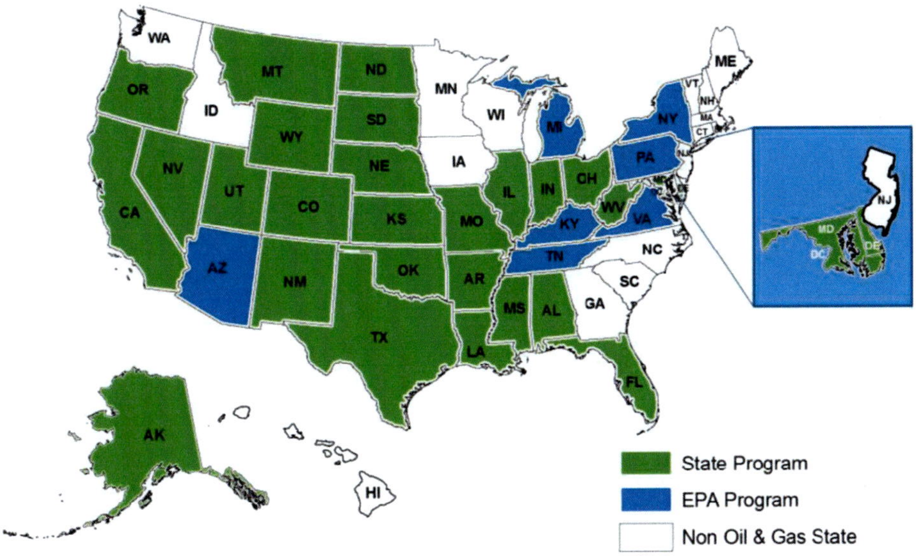

Source: EPA, 2008

Exhibit 27. UIC Class II Primacy Map

To date, 40 states have obtained primacy for oil and gas injection wells (Class II), although, as shown in Exhibit 27 not all of these states have oil and gas production. The U.S. EPA administers UIC programs for ten states, seven of which are oil and gas states, and all other federal jurisdictions and Indian Lands[183] (Exhibit 27[184]).

Oil Pollution Act of 1990 – Spill Prevention Control and Countermeasure

The CWA and the Oil Pollution Act (OPA) include both regulatory and liability provisions that are designed to reduce damage to natural resources from oil spills. Congress added Section 311 to the CWA, which in part authorized the President to issue regulations establishing procedures, methods, equipment, and other requirements to prevent discharges of oil from vessels and facilities [Section 311(j)(1)(c)]. In response to the Exxon Valdez oil spill in Alaska, Congress enacted the OPA in 1990[185]. The OPA amended CWA Section 311 and contains provisions applicable to onshore facilities and operations.

Section 311, as amended by the OPA, provides for spill prevention requirements, spill reporting obligations, and spill response planning. It regulates the prevention of and response to accidental releases of oil and hazardous substances into navigable waters, on adjoining shorelines, or affecting natural resources belonging to or managed by the U.S. This authority is primarily carried out through the creation and implementation of facility and response plans. These plans are intended to establish measures that will prevent discharge of oil into navigable waters of the U.S. or adjoining shore-lines as opposed to response and cleanup after a spill occurs.

A cornerstone of the strategy to prevent oil spills from reaching the nation's waters is the oil Spill Prevention, Control and Countermeasure (SPCC) plan. EPA promulgated regulations to implement this part of the OPA of 1990. These regulations specify that:

1. SPCC Plans must be prepared, certified (by a professional engineer) and implemented by facilities that store, process, transfer, distribute, use, drill for, produce, or refine oil;
2. Facilities must establish procedures and methods and install proper equipment to prevent an oil release;
3. Facilities must train personnel to properly respond to an oil spill by conducting drills and training sessions; and,
4. Facilities must also have a plan that outlines steps to contain, clean up and mitigate any effects that an oil spill may have on waterways[186].

Before a facility is subject to the SPCC rule, it must meet three criteria:

1. It must be non-transportation-related;
2. It must have an aggregate aboveground storage capacity greater than 1,320 gallons (31.4 bbls) or a completely buried storage capacity greater than 42,000 gallons (1,000 bbls); and
3. There must be a reasonable expectation of a discharge into or upon navigable waters of the U.S. or adjoining shorelines.

An SPCC Plan is a site-specific document that describes the measures the facility owner has taken to prevent oil spills, and what measures are in place, if needed, to contain and clean spills. It includes information about the facility, the oil storage containment, inspections, and a site diagram with locations of tanks (above and below ground) and drainage, and other pertinent details. Prevention measures include secondary containment around tanks and certain oil-containing equipment.

The SPCC program is not as applicable to shale gas operations as it is to oil production sites. Shale gas operators may have to prepare plans if they store large amounts of fuel (exceeding the volumes stated above) on site, or if oil-filled equipment is present, and there is a risk of that oil impacting waters of the U.S.

In October 2007, EPA proposed amendments to the SPCC rule intended to increase clarity and tailor certain requirements to ensure increased compliance. Among other things, these amendments would streamline some requirements by allowing the use of a plan template for smaller facilities, extending some deadlines for plan preparation, and exempting some vessels and flow lines from secondary containment requirements. They would also add spill prevention requirements for some oil and gas facilities. These proposed rules have not yet been made final[187].

State Regulations and Regional Cooperation

In addition to implementing federal statutes for the NPDES, UIC, and storm water programs, states and tribes may impose their own requirements to protect their water resources, both surface and underground. For example, they establish water quality standards for some or all of their surface water. These standards are approved by EPA and become the baseline for CWA permits[188].

In addition, some areas have established regional water authorities that regulate water withdrawals and discharges within a river basin. For example, the Susquehanna River Basin Commission (SRBC)[189] and the DRBC[190] in New York and Pennsylvania require that entities seeking to withdraw water from their river systems first obtain permits. These commissions have authority separate from the states. They have recently directed their attention to the water requirements for drilling and hydraulically fracturing Marcellus Shale gas wells and are updating their requirements for both water withdrawals and discharge of the water after use. Other river basin commissions are more advisory in nature, providing water flow and quality information and coordinating river conservation efforts by state agencies and others.

State agencies are the principal organizations for enforcing water quality regulations. They have inspectors, usually located at regional offices throughout the state, who visit oil and gas well sites to ensure compliance with regulations. When a violation occurs, state enforcement and legal personnel develop the case and order compliance, in many cases also imposing penalties against the violator. Penalties can range from fines to revocation of permits, and even to criminal sanctions in severe cases. Such penalties are usually imposed only after hearings before a board of commissioners or other state body. In addition to fines and penalties, companies that pollute surface or ground water must clean up or remediate the contamination they caused.

Regulation of Impacts on Air Quality

Air quality impacts are regulated under the Clean Air Act (CAA). As described below, the Act sets national standards for emissions of certain pollutants and requires permits for some industrial operations. Greenhouse Gases are not regulated as such, and are not, therefore, discussed in this section.

Clean Air Act

The CAA is the primary means by which EPA regulates potential emissions that could affect air quality. The U.S. Congress passed the CAA in 1963, and they have amended it on several occasions since, most recently in 1990[191]. The CAA requires EPA to set national standards to limit levels of certain pollutants. EPA regulates those pollutants by developing human health-based and/or environmentally and scientifically based criteria for setting permissible levels. Air regulations do not normally include exceptions for a company's size, the age of a field, or the type of operation. Typically, the air rules are silent on issues such as conventional versus unconventional plays, old versus new fields, and the depth of a well. For the most part, the air emissions, applicable regulations, and associated emissions controls for a shale play are no different than those for any other natural gas operation. There may be differences due to location (some areas of the country have better air quality than others), equipment needs (some shale plays may produce a wetter gas than others), and sulfur content level of the gas.

Geographic areas that do not meet EPA's standards for a given pollutant are designated as "nonattainment areas"[192]. This is the case for the Barnett Shale play, much of which is located in or near the Dallas-Fort Worth ozone nonattainment area. As a result, Barnett Shale production activities must often comply with much more stringent regulations than similar operations proposed outside of a nonattainment area. As a result of the implementation of the CAA, air quality has improved dramatically across the U.S. during the last few decades and existing regulations should continue to reduce air pollution emissions during the next twenty years or longer[193].

Air Quality Regulations

Like any other U.S. industry, shale gas producers must comply with existing and new air regulations including those resulting from the 1990 CAA Amendments. These rules pose an ongoing challenge to company resources as producers strive to understand and comply with enforcement, fines, public reaction, and possibly even project cancellations in light of new standards.

EPA has established National Emission Standards for Hazardous Air Pollutants (NESHAPs), which are nationally uniform standards to control specific air emissions. In 2007, EPA implemented a new standard referred to as the Maximum Achievable Control Technology (MACT) standard for hazardous air pollutants (HAP) that targeted small area sources such as shale gas operations located in areas near larger populations. These standards limit HAP emissions (primarily benzene) from process vents on glycol dehydration units, storage vessels with flash emissions, and equipment leaks.

Another example of new or amended federal regulations that will have a direct impact on controlling emissions from shale gas operations is the Stationary Spark Ignition Internal Combustion Engine new source performance standard[194] and Reciprocating Internal Combustion Engine NESHAP[195] rules, which regulate new and refurbished engines. These rules, passed in 2007, target all internal combustion engines regardless of horsepower rating, location, or fuel (electric engines are not included) and include extensive maintenance, testing, monitoring, recordkeeping, and reporting requirements[196].

EPA is not large enough to regulate every air emissions source nationwide, let alone consider the local and regional differences. Therefore, they typically delegate that role to local, state, and tribal agencies. This delegation of authority can include rule implementation, permitting, reporting, and compliance. Any state given such delegation of authority can pass more restrictive rules, but they are prohibited from passing a rule that is less stringent than its federal counterpart.

Air Permits

Air permits are legal documents that facility owners and operators must abide by. The permit specifies what construction is allowed, what emission limits must be met, how the emissions source(s) must be operated, and what conditions—specifying monitoring, record keeping, and reporting requirements—must be maintained to assure ongoing compliance. Shale gas producers may need air quality permits for a number of emissions sources, including gas compressor engines, glycol dehydrators, and flares.

A company's permitting responsibility does not end with the issuance of their initial air permit. They must be constantly vigilant that a new regulation, modification, replacement, or process change does not impact their existing permit and require a permit amendment or a more stringent permit. Although these permits may differ across the country, they all contain specific conditions designed to ensure state and federal standards are met and to prevent any significant degradation in air quality as a result of a proposed activity.

Regulation of Impacts to Land

Impacts to land from shale gas operations include solid waste disposal and surface disturbances that may impact the visual landscape or may affect wildlife habitat. Operations on federal lands are a special case with unique requirements that are discussed below.

Resource Conservation and Recovery Act (RCRA)

RCRA was passed in 1976 to address the growing problems of the increasing volume of municipal and industrial waste. RCRA established goals for protecting human health and the environment, conserving resources, and reducing the amount of waste. RCRA Subtitle C established a federal program to manage hazardous wastes from cradle to grave to ensure that hazardous waste is handled in a manner that protects human health and the environment. Subtitle D of the RCRA addresses non-hazardous solid wastes, including certain hazardous wastes which are exempted from the Subtitle C regulations[197].

In 1978, EPA proposed hazardous waste management standards that included reduced requirements for some industries, including oil and gas, with large volumes of wastes. EPA determined that these large volume "special wastes" were lower in toxicity than other wastes being regulated as hazardous waste under the RCRA[198].

In 1980, the Solid Waste Disposal Act (SWDA) amended RCRA to exempt drilling fluids, produced waters, and other wastes associated with exploration, development, and production of crude oil, natural gas and geothermal energy[199]. The SWDA Amendments also required EPA to provide a report to Congress on these wastes and to make a regulatory determination as to whether regulation of these wastes under RCRA Subtitle C was warranted[200].

In 1988, EPA issued a final regulatory determination stating that control of oil and gas exploration and production wastes under RCRA Subtitle C was not warranted.

In 1987, EPA issued a Report to Congress that outlined the results of a study on the management, volume, and toxicity of wastes generated by the oil, natural gas and geothermal industries. In 1988, EPA issued a final regulatory determination stating that control of oil and gas exploration and production wastes under RCRA Subtitle C was not warranted. EPA made this determination because it found that other state and federal programs could protect human health and the environment more effectively. In lieu of regulation under Subtitle C, EPA implemented a three-pronged strategy to ensure that the environmental and programmatic issues were addressed:

1. Improve other federal programs under existing authorities;
2. Work with states to improve some programs; and
3. Work with Congress to develop any additional statutory authorities that may be required[201].

These wastes have remained exempt from Subtitle C regulations, but this does not preclude these wastes from control under state regulations or other federal regulations[202]. The exemption applies only to the federal requirements of RCRA Subtitle C. A waste that is exempt from Subtitle C regulation might be subject to more stringent or broader state hazardous and non-hazardous waste regulations and other state and federal program regulations. For example, oil and gas exploration and production wastes may be subject to regulation under RCRA Subtitle D, the Clean Air Act, the Clean Water Act, the Safe Drinking Water Act, and/or the Oil Pollution Act of 1990[203,204].

In 1989, EPA worked with the Interstate Oil and Gas Compact Commission (IOGCC), state regulatory officials, industry representatives, and nationally recognized environmental groups to establish a Council on Regulatory Needs. The purpose of the council was to review existing state oil and gas exploration and production waste management programs and to develop guidelines to describe the elements necessary for an effective state program. This effort was begun by EPA as part of the second prong of the agency's approach. These groups then worked together with state regulatory agencies to review the state programs, on a voluntary basis, against these guidelines and to make recommendations for improvement. This state review program continues today under the guidance of a non-profit organization called STRONGER. The state programs reviewed to date represent over 90% of the onshore domestic production[205].

Working with the IOGCC, STRONGER has continued to update the guidelines consistent with developing environmental and oilfield technologies and practices. Under the state review process, state programs have continued to improve, and follow-up reviews have shown significant improvement where states have successfully implemented the recommendations of the review committees.

Endangered Species Act

The Endangered Species Act (ESA) of 1973 (Pub. L. 93-205) protects plants and animals that are listed by the federal government as "endangered" or "threatened"[206]. Sections 7 and 9 are central to regulating oil and gas activities. Section 9 makes it unlawful for anyone to "take" a listed animal, and this includes significantly modifying its habitat[207]. This applies to private parties and private land; a landowner is not allowed to harm an endangered animal or its habitat on his or her property.

Section 7 applies not to private parties, but to federal agencies. This section covers not only federal activities but also the issuance of federal permits for private activities, such as Section 404 permits issued by the Corps of Engineers, to people who want to do construction work in waters or wetlands[208]. Section 7 imposes an affirmative duty on federal agencies to ensure that their actions (including permitting) are not likely to jeopardize the continued existence of a listed species (plant or animal) or result in the destruction or modification of critical habitat. Both Sections 7 and 9 allow "incidental takes" of threatened or endangered species, but only with a permit.

To "take" is to harass, harm, pursue, hunt, shoot, wound, kill, trap, capture, or collect a plant or animal of any threatened or endangered species. Harm includes significant habitat modification when it kills or injures a member of a listed species through impairment of essential behavior (e.g., nesting or reproduction).

For any non-federal industrial activity, the burden is on the owner and/or operator to determine if an incidental take permit is needed. This is typically accomplished by contacting the U.S. Fish and Wildlife Service (FWS) to determine whether any listed species are present or will potentially inhabit the project site. A biological survey may be required to determine whether protected species are present on the site and whether a Section 9 permit may be required[209,210]. The FWS as well as many state fish and game agencies offer services to help operators determine whether a given project is likely to result in a take and whether a permit is required. FWS can also provide technical assistance to help design a project so as to avoid impacts. For example, the project could be designed to minimize disturbances during nesting or mating seasons[211].

A Section 9 permit must include a habitat conservation plan (HCP) consisting of: an assessment of impacts; measures that will be undertaken to monitor, minimize and mitigate any impacts; alternative actions considered and an explanation of why they were not taken; and any additional measures that the FWS may require[212]. Mitigation measures, which are actions that reduce or address potential adverse effects of a proposed activity upon species, must be designed to address the specific needs of the species involved and be manageable and enforceable. Mitigation measures may take many forms, such as preservation (via acquisition or conservation easement) of existing habitat; enhancement or restoration of degraded or former habitat; creation of new habitats; establishment of buffer areas around existing habitats; modifications of land use practices; and restrictions on access[213].

State Endangered Species Protections

All fifty states have fish and game/wildlife agencies that work in cooperation with the U. S. FWS district offices with regard to the incidental take permitting process. Many states also have their own endangered and threatened species lists that may include species not on the federal lists, and have their own requirements for protecting endangered species[214].

Oil and Gas Operations on Public Lands

Federal Lands

The U.S. Department of Interior's Bureau of Land Management (BLM) is responsible for permitting and managing most onshore oil and gas activities on federal lands. The BLM carries out its responsibility to protect the environment throughout the process of oil and gas resource exploration and development on public lands. Resource protection is considered throughout the land use planning process—when Resource Management Plans (RMPs) are prepared and when an Application for Permit to Drill (APD) is processed[215]. The BLM's inspection and enforcement and monitoring program is designed to ensure that operators comply with relevant laws and regulations as well as specific stipulations set forth during the permitting process.

Since most shale gas activity in the near future is expected to occur in the eastern U.S. basins, it is not likely that much of this development will occur on federal lands. While there are some federal lands, such as National Parks, National Forests, and military installations, these are much less extensive in the east than in the west. Where shale gas operations do occur on federal lands, BLM has a well established program for managing these activities to protect human health and the environment.

State Lands

The amount of state-owned land varies considerably from state to state and each state manages these lands differently. In most states, leasing of state-owned minerals occurs through lease auctions. Since states are already set up to manage oil and gas operations within their borders, no special permitting or enforcement systems are required. Some states do have Environmental Policy Acts that require a review of environmental impacts that may result from leasing or operations on state lands or of any state action that may affect the environment.

Other Federal Laws and Requirements that Protect the Environment

Comprehensive Environmental Response, Compensation, and Liability Act

The Comprehensive Environmental Response, Compensation, and Liability Act (CERCLA), commonly known as Superfund, was enacted by Congress on December 11, 1980. This law created a tax on the chemical and petroleum industries and provided broad federal authority to respond directly to releases or threatened releases of hazardous substances that may endanger public health or the environment. CERCLA established prohibitions and requirements concerning closed and abandoned hazardous waste sites, provided for liability of persons responsible for releases of hazardous waste at these sites, and established a trust fund to provide for cleanup when no responsible party could be identified. Over five years, $1.6 billion was collected and placed in a trust fund for cleaning up abandoned or uncontrolled hazardous waste sites.

CERCLA was amended by the Superfund Amendments and Reauthorization Act (SARA) in 1986. SARA made several changes to the Superfund program that augmented its effectiveness, provided new enforcement authorities, boosted state and citizen involvement, and increased the size of the trust fund.

In addition to the provisions for cleaning up hazardous waste sites, CERCLA requires the person in charge of a vessel or facility to immediately notify the National Response Center when there is a release of a hazardous substance in an amount equal to or greater than the reportable quantity (RQ) for that substance [CERCLA Section 103(a)]. The reportable quantity depends on the substance released.

CERCLA Section 101(14) excludes certain substances from the definition of hazardous substance, thus exempting them from CERCLA regulation. These substances include petroleum, meaning crude oil or any fraction thereof that is not specifically listed as a hazardous substance, natural gas, natural gas liquids, liquefied natural gas, and synthetic gas usable for fuel. If a release of one of these substances occurs, CERCLA notification is not required. Thus, CERCLA reporting will only apply to shale gas production and processing sites if hazardous substances other than crude oil or natural gas are spilled in reportable quantities; such are not usually present at these sites.

However, this particular exclusion applies only to CERCLA Section 103(a) reporting requirements; it does not exempt a facility from the Emergency Planning and Community Right-to-Know Act (EPCRA) Section 304 reporting requirements. A release of a petroleum product containing certain substances is potentially reportable under EPCRA Section 304 if more than an RQ of that substance is released[216].

Many states have separate requirements regarding hazardous substances. Reporting of releases of the materials exempted under CERCLA may be required under state law.

Emergency Planning and Community Right-to-Know Act

Congress enacted EPCRA in 1986 to establish requirements for federal, state and local governments, tribes, and industry regarding emergency planning and "community right-to-know" reporting on hazardous and toxic chemicals. The community right-to-know provisions of EPCRA are the most relevant part of the law for shale gas producers. They help increase the public's knowledge and access to information on chemicals at individual facilities, along with their uses and potential releases into the environment.

Under Sections 311 and 312 of EPCRA, facilities manufacturing, processing, or storing designated hazardous chemicals must make Material Safety Data Sheets (MSDSs), describing the properties and health effects of these chemicals, available to state and local officials and local fire departments. Facilities must also provide state and local officials and local fire departments with inventories of all on-site chemicals for which MSDSs exist. Information about chemical inventories at facilities and MSDSs must be available to the public. Facilities that store over 10,000 pounds of hazardous chemicals are subject to this requirement. Any hazardous chemicals above the threshold stored at shale gas production and processing sites must be reported in this manner.

Section 313 of EPCRA authorizes EPA's Toxics Release Inventory (TRI), which is a publicly available database that contains information on toxic chemical releases and waste management activities reported annually by certain industries as well as federal facilities. EPA issues a list of industries that must report releases for the database. To date, EPA has not included oil and gas extraction as an industry that must report under TRI. This is not an exemption in the law. Rather it is a decision by EPA that this industry is not a high priority for reporting under TRI. Part of the rationale for this decision is based on the fact that most of the information required under TRI is already reported by producers to state agencies that make it publicly available. Also, TRI reporting from the hundreds of thousands of oil and gas sites would overwhelm the existing EPA reporting system and make it difficult to extract meaningful data from the massive amount of information submitted[217, 218].

EPCRA section 304 requires reporting of releases to the environment of certain materials that are subject to this law. As noted in the section above, this requirement would apply to any releases of petroleum products that exceed reporting thresholds, even if those products are exempt from CERCLA reporting. While shale gas production facilities do not normally store the materials subject to EPCRA reporting, known as EPCRA "Extremely Hazardous Substances" and CERCLA hazardous substances, a limited number of chemicals used in the hydraulic fracturing process, such as hydrochloric acid, are classified as hazardous under CERCLA. These chemicals may be brought on site for a few days, at most, during fracturing or work-over operations. Businesses must report non-permitted releases—into the atmosphere, surface water, or groundwater—of any listed chemical above threshold amounts, known as the "reportable quantity", to federal, state, and local authorities. Therefore, while every precaution is taken to prevent chemical spills, in the event of an accidental release above the reportable quantity, a report would be made to these authorities by the operator.

Occupational Safety and Health Act

Under the Occupational Safety and Health Act of 1970, employers are responsible for providing a safe and healthy workplace for their employees. The Occupational Safety and Health Administration (OSHA) promotes the safety and health of America's working men and women by setting and enforcing standards; providing training, outreach and education; establishing partnerships; and encouraging continual process improvement in workplace safety and health[219].

OSHA has developed specific standards to reduce potential safety and health hazards in the oil and gas drilling, servicing and storage industry[220]. States also have requirements that provide further worker and public safety protections.

Summary

The U.S. has a long history of actively regulating the oil and gas industry including the shale gas industry. A comprehensive set of federal and state laws and programs regulate all aspects of shale gas exploration and production activities. Under these programs, federal, state and local agencies enforce an array of requirements designed to protect human health and the environment during drilling, production, and abandonment operations. Together, these requirements have reduced environmental risk and adverse impacts to our water, air, and land nationwide.

A comprehensive set of federal and state laws and programs regulate all aspects of shale gas exploration and production activities.

Environmental Considerations

As described in the previous sections, natural gas is an important part of the nation's energy supply. As a clean-burning, affordable and reliable source of energy, natural gas will continue to play a significant role in the energy supply picture for years to come. Unconventional sources of natural gas have become a major component of that future supply and shale gas is rapidly emerging as a critical part of that resource.

There exists an extensive framework of federal, state, and local requirements designed to manage virtually every aspect of the natural gas development process. These regulatory efforts are primarily led by state agencies and include such things as ensuring conservation of gas resources, prevention of waste, and protection of the rights of both surface and mineral owners while protecting the environment[221]. As part of their environmental protection mission, state agencies are responsible for safeguarding public and private water supplies, preserving air quality, addressing safety, and ensuring that wastes from drilling and production are properly contained and disposed of[222].

In order to make sound decisions about future shale gas development, it is important to understand the process of drilling and producing shale gas wells (Exhibit 28) and the attendant environmental considerations. A key element in the emergence of shale gas production has been the refinement of cost-effective horizontal drilling and hydraulic fracturing technologies. These two processes, along with the implementation of protective BMPs, have allowed shale gas development to move into areas that previously would have been inaccessible. Accordingly, it is important to understand the technologies and practices employed by the industry and their ability to prevent or minimize the potential effects of shale gas development on human health and the environment and on the quality of life in the communities in which shale gas production is located.

Many of the human and environmental considerations associated with shale gas production are common to all oil and gas development. However, the horizontal drilling and hydraulic fracturing that have become the standard for modern shale gas development bring with them new considerations as well as new ways to reduce impacts. As shale gas development has spread into more densely populated areas, new challenges have been encountered and new technologies and practices have been developed to meet these challenges. In addition, collaborations between industry, regulators and the public have created innovative environmental solutions to problems that at first seemed insurmountable.

Collaborations between industry, regulators and the public have created innovative environmental solutions to problems that at first seemed insurmountable.

One consideration associated with traditional gas development has been the surface disturbance required for access roads and well pads. As described in greater detail below, horizontal drilling provides a means to significantly reduce surface disturbance and a host of related concerns.

Another set of considerations associated with traditional oil and gas development are the conflicts that arise from split estates. In some instances mineral rights and surface rights are not owned by the same party. This is referred to as “split estate” or "severed minerals". The condition of split estate is more prevalent in western states where the federal government owns much of the mineral rights[223]. In the mid-west and eastern states, where shale gas development resources are more prevalent, only 4% of the lands are associated with a federal split estate[224]. However, these same areas frequently have private-private split estate scenarios where the surface owner differs from the mineral estate owner. In these cases the mineral owner may be another individual or a business enterprise such as a coal company.

It is important to understand that surface owners who do not own minerals rights are still afforded certain protections.

Mineral Leasing

Companies negotiate a private contract or lease that allows mineral development and compensates the mineral owners. Lease terms vary and can contain stipulations or mitigation measures pertinent to protect various resources. (Several weeks to years)

Permits

The operator must obtain a permit authorizing the drilling of a new well. Surveys, drilling plans, and other technical information are frequently required for a permit application. The approved permit may require site specific environmental protection measures. Other permits such as water withdrawal or injection permits may also be required. (Several weeks to months)

Road and Pad Construction

Once permits are received, roads are constructed to access the wellsite. Well pads are constructed to safely locate the drilling rig and associated equipment during the drilling process. Pits may be excavated to contain drilling fluids. (Several days to weeks)

Drilling and Completion

A drilling rig drills the well and multiple layers of steel pipe (called casing) are put into the hole and cemented in place to protect fresh water formations. (Weeks or months)

Hydraulic Fracturing

A specially designed fracturing fluid is pumped under high pressure into the shale formation. The fluid consists primarily of water along with a proppant (usually sand) and about 2% or less of chemical additives. This process creates fractures in rock deep underground that are "propped" open by the sand, which allows the natural gas to flow into the well. (Days)

Production

Once the well is placed on production, parts of the wellpad that are no longer needed for future operations are reclaimed. The gas is brought up the well, treated to a useable condition, and sent to market. (Interim Reclamation: days; Production: years)

Workovers

Gas production usually declines over the years. Operators may perform a workover which is an operation to clean, repair and maintain the well for the purposes of increasing or restoring production. Multiple workovers may be performed over the life of a well. (Several days to weeks)

Plugging and Abandonment/Reclamation

Once a well reaches its economic limit, it is plugged and abandoned according to State standards. The disturbed areas, including well pads and access roads, are reclaimed back to the native vegetation and contours or to conditions requested by the surface owner. (Reclamation Activity: Days; Full Restoration: Years)

Exhibit 28. Process of Shale Gas Development (Duration)

A split-estate situation, regardless of its nature, can result in conflicts—especially in areas where active mineral resource development is not commonplace. Land-owners can be surprised to find that the mineral lease holder is entitled to reasonable use of the land surface even though they do not own the surface. However, it is important to understand that surface owners who do not own minerals rights are still afforded certain protections. If the mineral owner does not own the surface where drilling will occur, a separate agreement may be negotiated (in some states it is required) with the land owner to ensure that he or she is compensated for the use of the land and to set requirements for reclaiming the land when operations are complete[225].

Shale gas development within or near existing communities has created challenges for production companies. New technologies have generally allowed these challenges to be met successfully. In some cases, a combination of modern shale gas technologies and the innovative use of BMPs has been required to allow development to continue without compromising highly valued community resources.

In one instance, Chesapeake Energy Corporation constructed a well pad near a popular Fort Worth community area, known as the Trinity Trail System, to develop natural gas from the Barnett Shale. The Trinity Trail System is located on private land and consists of a 35-mile network of paved and natural surface pathways. The drilling pad was constructed approximately 200 feet from one portion of the trail. During the initial planning stages, proposed use of this land for development of natural gas was met with significant opposition by the public. Maintaining healthy populations of upland hardwood forest habitat was important to the community because such woodlots are rare in urban settings. To address the concerns of the community, the company sponsored public meetings and opinion surveys; provided landscape plans; planted trees and shrubs; and enhanced the general area by improving irrigation and lowering maintenance requirements. The well pad was specifically designed to be as small as possible in order to reduce the well's footprint. Preventative construction practices were used that helped to preserve many of the existing trees. The construction zone was isolated from view using a 16-ft barrier fence with sound baffling. This approach benefitted both parties: the company was able to produce the shale gas, important community resources were protected, and at no point in the process was any portion of the trail closed[226].

The following discussions describe the general process of development with emphasis on the horizontal drilling and hydraulic fracturing technologies that are the hallmarks of modern shale gas production. The section also describes the environmental considerations that accompany shale gas development and the technologies and practices that are in place to prevent or minimize impacts.

Horizontal Wells

Modern shale gas development is a technologically driven process for the production of natural gas resources. Currently, the drilling and completion of shale gas wells includes both vertical and horizontal wells (Exhibit 29). The emerging shale gas basins are expected to follow a trend similar to the Barnett Shale play with increasing numbers of horizontal wells as the plays mature[227,228,229]. The technologies utilized by operators to drill shale gas wells are similar to the drilling techniques that have been industry standards for drilling of conventional gas wells. Both horizontal drilling and hydraulic fracturing are established technologies with significant track records; horizontal drilling dates back to the 1930s and hydraulic fracturing has a history dating back to the 1950s[230]. The key difference between a shale gas well and a conventional gas well is the reservoir stimulation (large-scale hydraulic fracturing) approach performed on shale gas wells[231].

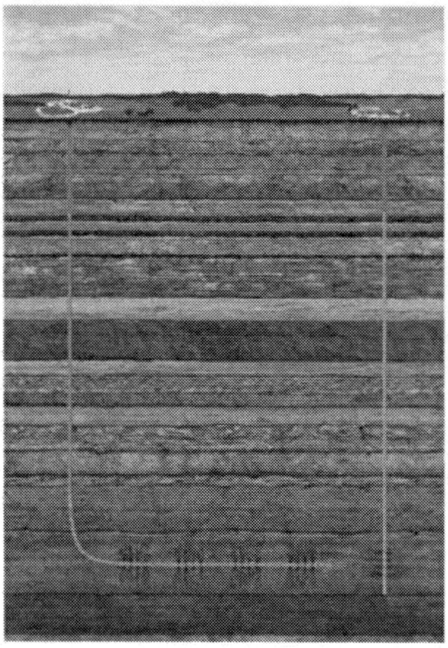

Source: John Perez, Copyright ©, 2008

Exhibit 29. Horizontal and Vertical Well Completions

The evolution of the Barnett Shale play toward favoring horizontal wells resulted from improvements in the technology combined with the economic benefits of the greater reservoir exposure that a horizontal well provides over a vertical well. While both well types may be used to recover the resource, shale gas operators are increasingly relying on horizontal well completions to optimize recovery and well economics[232]. Exhibit 29 illustrates how horizontal drilling provides more exposure to a formation than does a vertical well. For example, in the Marcellus Shale in Pennsylvania, a vertical well may be exposed to as little as 50 ft of formation while a horizontal well may have a lateral wellbore extending in length from 2,000 to 6,000 ft within the 50- to 300-ft thick formation[233]. This increase in reservoir exposure creates a number of advantages over vertical wells drilling.

Both horizontal drilling and hydraulic fracturing are established technologies with significant track records; horizontal drilling dates back to the 1930s and hydraulic fracturing has a history dating back to the 1950s.

There are a wide range of factors that influence the choice between a vertical or horizontal well. While vertical wells may require less capital investment on a per well basis, production is often less economical. A vertical well may cost as much as $800,000 (excluding pad and infrastructure) to drill compared to a horizontal well that can cost $2.5 million or more (excluding pad and infrastructure)[234].

Reducing Surface Disturbance

Complete development of a 1-square mile section could require 16 vertical wells each located on a separate well pad. Alternatively, six to eight horizontal wells (potentially more), drilled from only one well pad, could access the same reservoir volume, or even more[235]. The low natural permeability of shale requires vertical wells to be developed at closer spacing intervals than conventional gas reservoirs in order to effectively manage the resource. This can result in initial development of vertical wells at spacing intervals of 40 acres per well, or less, to efficiently drain the gas resources from the tight shale reservoirs. In addition, horizontal drilling can significantly reduce the overall number of well pads, access roads, pipeline routes, and production facilities required, thus minimizing habitat fragmentation, impacts to the public, and the overall environmental footprint. Devon Energy Corporation reports that the use of horizontal wells in the Barnett Shale has allowed the company to replace 3 or 4 vertical wells with a single horizontal well. While it is too early to determine the final well spacing that will most efficiently recover the gas resource in all basins, experience to date indicates that the use of horizontal well technology will significantly decrease the total environmental disturbance.

Exhibit 11 includes data on well spacing for some of the developing shale gas basins. Using this data it is possible to compare the development of a typical 640-acre (1 square mile) area with vertical versus horizontal wells. The spacing interval for vertical wells in the gas shale plays averages 40 acres per well for initial development. The spacing interval for horizontal wells is likely to be approximately 160 acres per well. Therefore, a 640-acre section of land could be developed with a total of 16 vertical wells, each on its own individual well pad, or by as few as 4 horizontal wells all drilled from a single multi-well drilling pad. Analysis performed in 2008 for the U.S. Department of the Interior estimated that a shallow vertical gas well completed in the Fayetteville Shale in Arkansas would have a 2.0-acre well pad, 0.10 miles of road and 0.55 miles of utility corridor, resulting in a total of 4.8 acres of disturbance per well[236]. The same source identified a horizontal well pad in Arkansas as occupying approximately 3.5 acres plus roads and utilities, resulting in a total of 6.9 acres. If multiple horizontal wells are completed from a single well pad it may require the pad to be enlarged slightly. Estimating that this enlargement will result in a 0.5-acre increase, the 4-well horizontal pad with roads and utilities would disturb an estimated total of 7.4 acres, while the 16 vertical wells would disturb approximately 77 acres. In this example, 16 vertical wells would disturb more than 10 times the area of 4 horizontal wells to produce the same resource volume. This difference in development footprint when considered in terms of both rural and urban development scenarios highlights the desire for operators to move toward horizontal development of gas shale plays.

Reducing Wildlife Impacts

Research has documented that activities associated with gas development can affect wildlife and its habitat during the exploration, development, operations, and abandonment phases[237]. The development of shale gas utilizing horizontal wells and multi-well pads not only reduces surface area disturbances by reducing the total number of wells drilled and well pad sites constructed, but also results in fewer roadways and utility corridors. This overall reduction in a project's footprint results in significantly less habitat disturbance while allowing for more operational flexibility. Furthermore, by drilling underneath sensitive areas such as wetlands, areas near streams and rivers and wilderness habitats, gas can be produced without disturbing some of these resources. This ability to reduce surface disturbance is especially important in certain critical habitats. For example, certain portions of New York (e.g., Catskill Park, the Shawangunk Ridge, the Hudson Highlands and the Poconos) are dominated by hardwood forests, which are important wildlife habitats that are susceptible to fragmentation[238].

Source: ALL Consulting, 2008

Active Drilling Rig in the Barnett Shale Play

Source: WVSORO

Drilling Rig in Rural Upshur County, West Virginia

In addition, state regulations and, in some cases, local ordinances include stipulations dictating operational restrictions to provide added protection for wildlife or sensitive resources. In the city of Flower Mound, Texas, ordinances have been adopted to protect the surface resources and allow for future growth of the community without detracting from the land value or sense of community. These ordinances prevent construction in or near streams or rivers, floodplains and sensitive upland forest to protect wildlife species and their associated habitats.

At the state level, special plans or waivers are required when surface use actions may affect threatened or endangered species. Such waivers must demonstrate that contemplated disturbances will not adversely impact the species in question. In Pennsylvania, wildlife are further protected on state lands (by the Pennsylvania Game Commission) by using lease agreements that require, whenever feasible, the use of existing timber and maintenance roads to access wells and avoidance of areas such as wetlands and unique and critical habitats for threatened or endangered species[239].

When disturbances to wildlife habitat are unavoidable, energy companies mitigate land disturbances by implementing land reclamation practices to restore disturbed land to original conditions. In general, reclamation practices (or mitigation measures) designed to protect and maintain wildlife will depend on project features, regional characteristics, and the potentially affected species. However, because technologies associated with modern shale gas development can reduce impacts in the first place, the need for additional protective restoration measures may also be reduced. Regardless of the situation, the timely reclamation of disturbed lands (e.g., re-seeding, land contouring, and re-vegetating) can minimize short and long-term disturbances to natural habitats[240].

Reducing Community Impacts

States, local governments, and industry can work together in the initial planning phase of development to minimize long term effects and to address citizen concerns such as traffic congestion, damage to roads, dust, and noise [241].The process of shale gas development, especially drilling and hydraulic fracturing, can create short-term increases in traffic volume, dust and noise. These nuisance impacts are usually limited to the initial 20- to 30-day drilling and completion period[242]. Along with increases in traffic volume, damage to road surfaces can occur if design parameters for traffic volume and weight loads are exceeded. Where these effects are an issue, developers have worked with authorities to adjust work schedules to help alleviate congestion; water unpaved roads to reduce dust; and adjust timing of some operations and install special sound barriers to reduce noise for nearby residents. When feasible, developers can also use avoidance practices to help minimize traffic congestion on heavily traveled roads. In the Barnett Shale play around the Dallas-Fort Worth International Airport, operators have constructed permanent pipelines to transfer produced water from well sites to disposal facilities, thereby reducing traffic and potential damage to roads[243]. When these practices are coupled with the benefits of multiple directional wells from fewer pads, the number of access roads and associated traffic can be further reduced.

In many cases, developers have negotiated to compensate local municipalities for road damage that does occur as a result of their activities. Alternatively, they may negotiate road maintenance and repair agreements to ensure that damage to roadways are repaired and that the cost is absorbed by the drilling enterprises[244]. The Perryman Group, in their 2007 study of the Barnett Shale play, noted that although traffic volume is a legitimate concern in the area, developers were effectively addressing the issue through maintenance agreements so that road repairs do not adversely affect local taxpayers[245].

From a traffic perspective, members of the public or local municipalities often have the ability to limit traffic volume in residential areas by developing restrictions in neighborhood lease agreements or by developing ordinances that prevent road construction in certain areas, respectively. In urban areas these agreements can be used to coordinate local traffic patterns to minimize congestion, control speed limits to address safety concerns, and specify weight zones to reduce road damage.

Source: Parker County Commissioner's Office

Tanker Trucks in Parker County, Texas

Source: ALL Consulting, 2008

Shale Gas Activity at Dallas-Fort Worth International Airport

With continued advances in technologies, modern developers are afforded a higher level of drilling flexibility than in the past. This provides producers with the ability to adjust their operational plans allowing them to access drilling locations that would otherwise be inaccessible. Although drilling circumstances vary by geologic region and well location, in many cases, shale gas plays are being developed with both vertically and horizontally drilled wells (Exhibit 29). Based on the current development activities of active gas shale basins, horizontal drilling has become the preferred method of drilling in most shale gas plays. Horizontal wells have also been used in many areas of the country to remotely access natural gas resources beneath existing infrastructure, buildings, environmentally sensitive areas, or other features that would prevent the use of vertical wells. The development of the Barnett

Shale near Dallas-Fort Worth International Airport is a prime example of how development of urban areas is possible with horizontal wellbores[246].

Changes in practices during the drilling and completion of shale gas wells have evolved from the Barnett Shale play near Dallas-Fort Worth International Airport and other urban areas surrounding the airport. Development practices there have been altered to suit local ordinances implemented to lessen community impacts and protect environmental resources. These ordinances include detailed setbacks from residences, roadways, churches, and schools, and means to control visual and noise impacts including the required use of directional lighting[247]. This results in the use of BMPs for sound barriers and lighting. Typically, drilling operations in rural gas development areas continue around the clock until the well is completed. When these same operations moved into the urban areas around the cities of Arlington, Burleson, Cleburne, Fort Worth, Joshua and North Richland Hills, specific ordinances were developed requiring additional permitting, well set backs from properties, day-time and night-time noise limits, and directional lighting[248]. Directional lighting provides illumination of well sites for worker safety, directing the light downward and shielding the surrounding area to prevent illuminating neighboring residences, roads or other buildings[249].

The purpose of ordinances and best management practices is to facilitate the development of the natural gas resource while protecting quality of life and environmental values in the surrounding areas.

In a similar concept, these drilling rigs are also being outfitted with blanket-like enclosures that act as an acoustic barrier to reduce engine noise. Sound deadening technology is a BMP that is also being applied to reduce noises from compressor facilities in both rural and urban settings[250]. These sound barriers include developing alternative building materials with integral sound absorbing properties.

These "BMPs" are not appropriate for all operations and must be applied on a case-by-case basis. In some cases, a given BMP may actually be counter-productive. In other cases, a particular BMP may create other environmental, safety, or operational problems that must be weighed against each other. While BMPs have certain benefits in certain situations, they cannot be universally applied or required.

Protecting Groundwater: Casing and Cementing Programs

State oil and gas regulatory programs place great emphasis on protecting groundwater. Current well construction requirements consist of installing multiple layers of protective steel casing and cement that are specifically designed and installed to protect fresh water aquifers and to ensure that the producing zone is isolated from overlying formations. During the drilling process, a conductor and surface casing string are set in the borehole and cemented in place. In some instances, additional casing strings may be installed; these are known as intermediate casings (Exhibit 30[251]). After each string of casing is set, and prior to drilling any deeper in the borehole, the casing is cemented to ensure a seal is provided between the casing and formation or between two strings of casing[252]. Exhibit 30 illustrates the casing and cement that may be installed in shale gas wells and highlights how the casing can be set to isolate different water-bearing zones from each other. The exhibit shows the multiple strings of casing, layers of cement and the production tubing, which are all important parts of the well completion in preventing contamination of fresh water zones and assuring that the gas resource does not flow into other, lower pressure zones around the outside of the casing rather than flowing up the well to be produced and sold.[253.]

Source : Chesapeake Energy Corporation, 2008

Insulation Blankets Used to Deaden Noise from Drilling Operations

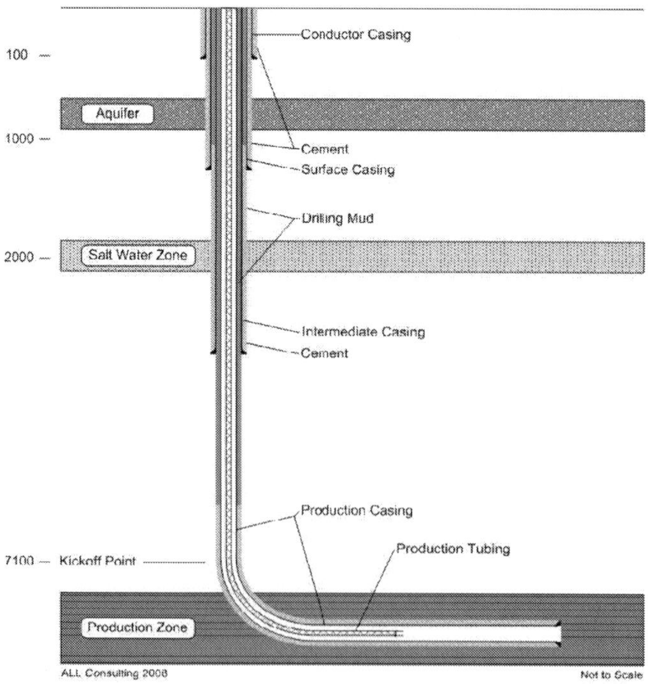

Exhibit 30. Casing Zones and Cement Programs

The conductor casing serves as a foundation for the well construction and prevents caving of surface soils. The surface casing is installed to seal off potential freshwater-bearing zones, this isolation is necessary in order to protect aquifers from drilling mud and produced fluids. As a further protection of the fresh water zones, air-rotary drilling is often used when drilling through this portion of the wellbore interval to ensure that no drilling mud comes in contact with the fresh water zone. Intermediate casings, when installed, are used to isolate non-freshwater-bearing zones from the producing wellbore. Intermediate casing may be necessary because of a naturally over-pressured zone or because of a saltwater zone located at depth. The borehole area below an intermediate casing may be uncemented until just above the kickoff point for the horizontal leg. This area of wellbore is typically filled with drilling muds.

Each string of casing serves as a layer of protection separating the fluids inside and outside of the casing and preventing each from contacting the other. Operators perform a variety of checks to ensure that the desired isolation of each zone is occurring including ensuring that the casing used has sufficient strength, and that the cement has properly bonded to the casing[254]. These checks may include acoustic cement bond logs and pressure testing to ensure the mechanical integrity of casings. Additionally, state oil and gas regulatory agencies often specify the required depth of protective casings and regulate the time that is required for cement to set prior to additional drilling. These requirements are typically based on regional conditions and are established for all wildcat wells and may be modified when field rules are designated. These requirements are instituted by the state oil and gas agency to provide protection of groundwater resources[255]. Once the casing strings are run and cemented there could be five or more layers or barriers between the inside of the production tubing and a water-bearing formation (fresh or salt).

Analysis of the redundant protections provided by casings and cements was presented in a series of reports and papers prepared for the American Petroleum Institute (API)[256] in the 1980s. These investigations evaluated the level of corrosion that occurred in Class II injection wells. Class II injection wells are used for the routine injection of water associated with oil and gas production. The research resulted in the development of a method of calculating the probability (or risk) that fluids injected into Class II injection wells could result in an impact to a USDW. This research started by evaluating data for oil and gas producing basins to determine if there were natural formation waters present that were reported to cause corrosion of well casings. The United States was divided into 50 basins, and each basin was ranked by its potential to have a casing leak resulting from such corrosion.

Detailed analysis was performed for those basins in which there was a possibility of casing corrosion[257]. Risk probability analysis provided an upper bound for the probability of the fracturing fluids reaching an underground source of drinking water. Based on the values calculated, a modern horizontal well completion in which 100% of the USDWs are protected by properly installed surface casings (and for geologic basins with a reasonable likelihood of corrosion), the probability that fluids injected at depth could impact a USDW would be between 2×10^{-5} (one well in 200,000) and 2×10^{-8} (one well in 200,000,000) if these wells were operated as injection wells. Other studies in the Williston basin found that the upper bound probability of injection water escaping the wellbore and reaching an underground source of drinking water is seven changes in one million well-years where surface casings cover the drinking water aquifers[258].

These values do not account for the differences between the operation of a shale gas well and the operation of an injection well. An injection well is constantly injecting fluid under pressure and thus raises the pressure of the receiving aquifer, increasing the chance of a leak or well failure. A production well is reducing the pressure in the producing zone by giving the gas and associated fluid a way out, making it less likely that they will try to find an alternative path that could contaminate a fresh water zone. Furthermore, a producing gas well would be less likely to experience a casing leak because it is operated at a reduced pressure compared to an injection well. It would be exposed to lesser volumes of potentially corrosive water flowing through the production tubing, and it would only be exposed to the pumping of fluids into the well during fracture stimulations.

The API study included an analysis of wells that had been in operation for many years when the study was performed in the late 1980s, and does not account for advances that have occurred in equipment and applied technologies and changes to the regulations. As such, a calculation of the probability of any fluids, including hydraulic fracture fluids, reaching a USDW from a gas well would indicate an even lower probability; perhaps by as much as two to three orders of magnitude. The API report came to another important conclusion relative to the probability of the contamination of a USDW when it stated that:

> ...for injected water to reach a USDW in the 19 identified basins of concern, a number of independent events must occur at the same time and go *undetected* [emphasis added]. These events include simultaneous leaks in the [production] tubing, production casing, [intermediate casing,] and the surface casing coupled with the unlikely occurrence of water moving long distances up the borehole past salt water aquifers to reach a USDW[259].

As indicated by the analysis conducted by API and others, the potential for groundwater to be impacted by injection is low. It is expected that the probability for treatable groundwater to be impacted by the pumping of fluids during hydraulic fracture treatments of newly installed, deep shale gas wells when a high level of monitoring is being performed would be even less than the 2 x 10-8 estimated by API.

In addition to the protections provided by multiple casings and cements, there are natural barriers in the rock strata that act as seals holding the gas in the target formation. Without such seals, gas and oil would naturally migrate to the earth's surface. A fundamental precept of oil and gas geology is that without an effective seal, gas and oil would not accumulate in a reservoir in the first place and so could never be tapped and produced in usable quantities. These sealing strata also act as barriers to vertical migration of fluids upward toward useable groundwater zones. Most shale gas wells (outside of those completed in the New Albany and the Antrim) are expected to be drilled at depths greater than 3,000 feet below the land surface (based on the data presented in Exhibit 11). Exhibit 31 compares estimated shallowest producible depth of the target ("pay") shale zone and the maximum base of treatable water. For any fluid present in the producing zone to reach treatable groundwater the fluid must migrate through these overlying zones.

A fundamental precept of oil and gas geology is that without an effective seal, gas and oil would not accumulate in a reservoir in the first place and so it could never be tapped and produced in usable quantities. These sealing strata also act as barriers to vertical migration of fluids upward toward groundwater zones.

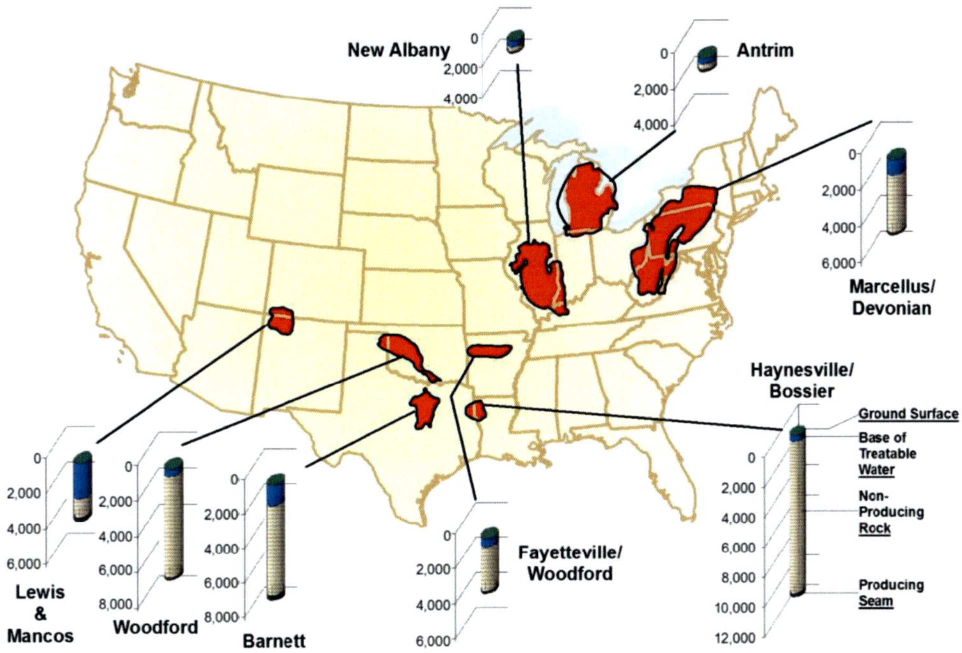

Source: Compiled from Various Data Sources

Exhibit 31.Comparison of Target Shale Depth and Base of Treatable Groundwater

Drilling Fluids and Retention Pits

Drilling fluids are a necessary component of the drilling process; they circulate cuttings (rock chips created as the drill bit advances through rock, much like sawdust) to the surface to clear the borehole, they lubricate and cool the drilling bit, they stabilize the wellbore (preventing cave in), and control downhole fluid pressure[260]. In order to maintain sufficient volumes of fluids onsite during drilling, operators typically use pits to store make-up water used as part of the drilling fluids. Storage pits are not used in every development situation. In the case of shale gas development, drilling operations have been occurring in both urban and rural locations, requiring that drilling practices be adapted to facilitate development in both settings. Drilling with compressed air is becoming an increasingly popular alternative to drilling with fluids due to the increased cost savings from both reduction in mud costs and the shortened drilling times as a result of air based drilling[261]. The air, like drilling mud, functions to lubricate, cool the bit, and remove cuttings. Air drilling is generally limited to low pressure formations, such as the Marcellus shale in New York[262].

In rural areas, storage pits may be used to hold fresh water for drilling and hydraulic fracturing. In an urban setting, due to space limitations, steel storage tanks may be used. Tanks can also be used in a closed-loop drilling system. Closed-loop drilling allows for the re-use of drilling fluids and the use of lesser amounts of drilling fluids[263]. Closed-loop drilling systems have also been used with water-based fluids in environmentally sensitive environments in combination with air-rotary drilling techniques[264]. While closed-loop drilling has been used to address specific situations, the practice is not necessary for every well drilled. As discussed in the previous section, drilling is a regulated practice managed at the state level, and while state oil and gas agencies have the ability to require operators to vary standard practices, the agencies typically do so only when it is necessary to protect the gas resources and the environment.

Source: ALL Consulting, 2008

Lined Fresh Water Supply Pit from the Marcellus Shale Development in Pennsylvania

In rural environments, storage pits may be used to hold water. They are typically excavated containment ponds that, based on the local conditions and regulatory requirements, may be lined. Pits can also be used to store additional make-up water for drilling fluids or to store water used in the hydraulic fracturing of wells.

Water storage pits used to hold water for hydraulic fracturing purposes are typically lined to minimize the loss of water from infiltration (notice the black synthetic liner in the accompanying photograph). Water storage pits are becoming an important tool in the shale gas industry because the drilling and hydraulic fracturing of these wells often requires significant volumes of water as the base fluid for both purposes[265].

Hydraulic Fracturing

The other technological key to the economic recovery of shale gas is hydraulic fracturing. Hydraulic fracturing is a formation stimulation practice used to create additional permeability in a producing formation, thus allowing gas to flow more readily toward the wellbore[266,267.] Hydraulic fracturing can be used to overcome natural barriers to the flow of fluids (gas or water) to the wellbore. Such barriers may include naturally low permeability common in shale formations or reduced permeability resulting from near wellbore damage during drilling activities[268].

Stimulations are optimized to ensure that fracture development is confined to the target formation.

Hydraulic fracturing involves the pumping of a fracturing fluid into a formation at a calculated, predetermined rate and pressure to generate fractures or cracks in the target formation. For shale gas development, fracture fluids are primarily water-based fluids mixed with additives which help the water to carry sand proppant into the fractures. The sand proppant is needed to "prop" open the fractures once the pumping of fluids has stopped. Once the fracture has initiated, additional fluids are pumped into the wellbore to continue the development of the fracture and to carry the proppant deeper into the formation. The additional fluids are needed to maintain the downhole pressure necessary to accommodate the increasing length of opened fracture in the formation. Each rock formation has inherent natural variability resulting in different fracture pressures for different formations. The process of designing hydraulic fracture treatments involves identifying properties of the target formation including fracture pressure, and the desired length of fractures. The following discussion addresses some of the processes involved in the design of a hydraulic fracture stimulation of a shale gas formation.

Fracture Design

Modern formation stimulation practices are sophisticated, engineered processes designed to emplace fracture networks in specific rock strata[269]. A hydraulic fracture treatment is a controlled process designed to the specific conditions of the target formation (thickness of shale, rock fracturing characteristics, etc.). Understanding the *in-situ* reservoir conditions present and their dynamics is critical to successful stimulations. Hydraulic fracturing designs are continually refined to optimize fracture networking and maximize gas production. While the concepts and general practices are similar, the details of a specific fracture operation can vary substantially from basin to basin and from well to well.

Fracture design can incorporate many sophisticated and state-of-the-art techniques to accomplish an effective, economic and highly successful fracture stimulation. Some of these techniques include modeling, microseismic fracture mapping, and tilt-meter analysis.

A computer model can be used to simulate hydraulic fracturing designs[270]. This approach helps maximize effectiveness and economically design a treatment event. The modeling programs allow geologists and engineers to modify the design of a hydraulic fracture treatment and evaluate the height, length, and orientation of potential fracture development (Exhibit 32)[271]. These simulators also allow the designers to use the data gathered during a fracture stimulation to evaluate the success of the fracture job performed. From these data and analyses, engineers can optimize the design of future fracture stimulations.

Additional advances in hydraulic fracturing design target analysis of hydraulic fracture treatments through technologies such as microseismic fracture mapping (Exhibit 33[272]) and tilt measurements[273]. These technologies can be used to define the success and orientation of the fractures created, thus providing the engineers with the ability to manage the resource through the strategic placement of additional wells, taking advantage of the natural reservoir conditions and expected fracture results in new wells.

Source: ALL Consulting, 2008

A Fracture Stimulation Is Closely Monitored by Many Specialists (Fayetteville Shale - Arkansas)

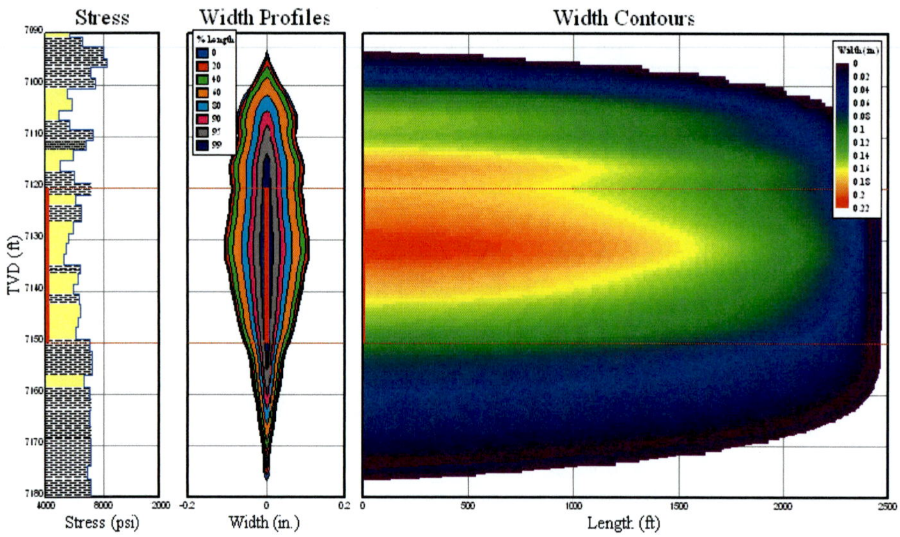

Source: Chesapeake, 2008

Exhibit 32.Example Output of a Hydraulic Fracture Stimulation Model

As more formation-specific data are gathered, service companies and operators can optimize fracture patterns. Operators have strong economic incentives to ensure that fractures do not propagate beyond the target formation and into adjacent rock strata[274]. Allowing the fractures to extend beyond the target formation would be a waste of materials, time, and money. In some cases, fracturing outside o f the target formation could potentially result in the loss of the well and the associated gas resource. Fracture growth outside of the target formation can result in excess water production from bounding strata. Having to pump and handle excess water increases production costs, negatively impacting well economics. This is a particular concern in the Barnett Shale of Texas where the underlying Ellenberger Group limestones are capable of yielding significant formation water.

Operators have strong economic incentives to ensure that fractures do not propagate beyond the target formation and into adjacent rock strata.

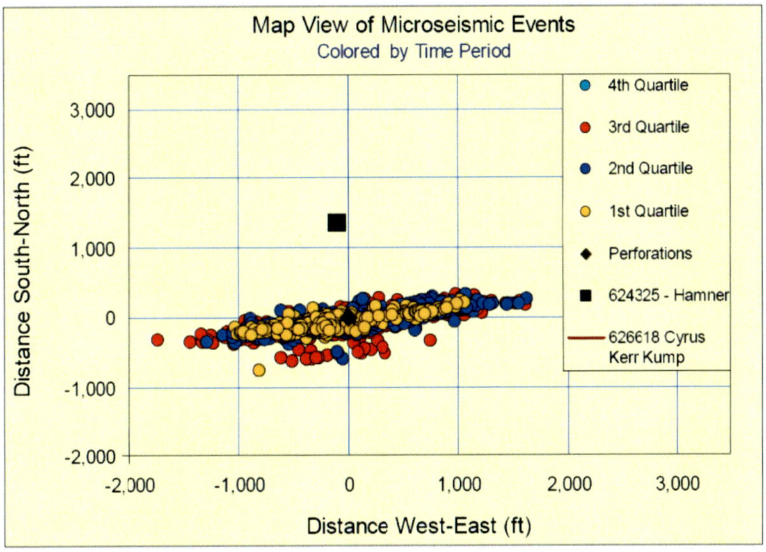

Source: Oilfield Service Company, 2008

Exhibit 33.Mapping of Microseismic Events

Fracturing Process

Hydraulic fracturing of horizontal shale gas wells is performed in stages. Lateral lengths in horizontal wells for shale gas development may range from 1,000 feet to more than 5,000 feet. Because of the length of exposed wellbore, it is usually not possible to maintain a downhole pressure sufficient to stimulate the entire length of a lateral in a single stimulation event[275]. Because of the lengths of the laterals, hydraulic fracture treatments of horizontal shale gas wells are usually performed by isolating smaller portions of the lateral. The fracturing of each portion of the lateral wellbore is called a stage. Stages are fractured sequentially beginning with the section at the farthest end of the wellbore, moving uphole as each stage of the treatment is completed until the entire lateral well has been stimulated[276]. Horizontal wells in the various gas shale basins may be treated using two or more stages to fracture the entire perforated interval of the well. Each stage of a horizontal well fracture treatment is similar to a fracture treatment for a vertical shale gas well.

For each stage of a fracture treatment, a series of different volumes of fracture fluids, called sub-stages, with specific additives and proppant concentrations, is injected sequentially. Exhibit 34[277] presents an example of the sub-stages of a single-stage hydraulic fracture treatment for a well completed in the Marcellus Shale[278]. This is a single-stage treatment typical of what might be performed on a vertical shale well or for each stage of a multi-stage horizontal well treatment. The total volume of the sub-stages in Exhibit 34 is

578,000 gallons. If this were one stage of a four-stage horizontal well, the entire fracture operation would require approximately four times this amount, or 2.3 million gallons of water.

Before operators or service companies perform a hydraulic fracture treatment of a well (vertical or horizontal), a series of tests is performed. These tests are designed to ensure that the well, well equipment and hydraulic fracturing equipment are in proper working order and will safely withstand the application of the fracture treatment pressures and pump flow rates. The tests start with the testing of well casings and cements during the drilling and well construction process. Testing continues with pressure testing of hydraulic fracturing equipment prior to the fracture treatment process[279]. It should be noted that construction requirements for wells are mandated by state oil and gas regulatory agencies to ensure that a well is protective of water resources and is safe for operation.

After the testing of equipment has been completed, the hydraulic fracture treatment process begins. The sub-stage sequence is usually initiated with the pumping of an acid treatment. This acid treatment helps to clean the near-wellbore area which can be “damaged” (pores and pore throats become plugged with drilling mud or casing cement) as a result of the drilling and well installation process. The next sequence after the acid treatment is a slickwater pad, which is a water-based fracturing fluid mixed with a friction reducing agent. The pad is a volume of fracturing fluid large enough to effectively fill the wellbore and the open formation area. The slickwater pad helps to facilitate the flow and placement of the proppant further into the fracture network.

After the pad is pumped, the first proppant sub-stage, combining a large volume of water with fine mesh sand is pumped. The next several sub-stagesin the stage increase the volume of fine-grained proppant while the volume of fluids pumped are decreased incrementally from 50,000 gallons (gals) to 30,000 gals. This fine-grained proppant is used because the finer particle size is capable of being carried deeper into the developed fractures[280]. In this example, the fine proppant sub-stages are followed by eight sub-stages of a coarser proppant with volumes from 20,000 gals to 10,000 gals. After the completion of the final sub-stage of coarse proppant, the well and equipment are flushed with a volume of freshwater sufficient to remove excess proppants from the equipment and the wellbore.

Hydraulic fracturing stimulations are overseen continuously by operators and service companies to evaluate and document the events of the treatment process. Every aspect of the fracture stimulation process is carefully monitored, from the wellhead and downhole pressures to pumping rates and density of the fracturing fluid slurry. The monitors also track the volumes of each additive and the water used, and ensure that equipment is functioning properly. For a 12,000-bbl (504,000-gallon) fracture treatment of a vertical shale gas well there may be between 30 and 35 people on site monitoring the entire stimulation process.

The staging of multiple fracture treatments along the length of the lateral leg of the horizontal well allows the fracturing process to be performed in a very controlled manner. By fracturing discrete intervals of the lateral wellbore, the operator is able to make changes to each portion of the completion zone to accommodate site-specific changes in the formation. These site-specific variations may include variations in shale thickness, presence or absence of natural fractures, proximity to another wellbore fracture system, and boreholes that are not centered in the formation.

Exhibit 34. Example of a Single Stage of a Sequenced Hydraulic Fracture Treatment

Hydraulic Fracture Treatment Sub-Stage	Volume (gallons)	Rate (gal/min)
Diluted Acid (15%)	5,000	500
Pad	100,000	3,000
Prop 1	50,000	3,000
Prop 2	50,000	3,000
Prop 3	40,000	3,000
Prop 4	40,000	3,000
Prop 5	40,000	3,000
Prop 6	30,000	3,000
Prop 7	30,000	3,000
Prop 8	20,000	3,000
Prop 9	20,000	3,000
Prop 10	20,000	3,000
Prop 11	20,000	3,000
Prop 12	20,000	3,000
Prop 13	20,000	3,000
Prop 14	10,000	3,000
Prop 15	10,000	3,000
Flush	13,000	3,000

Notes: Volumes are presented in gallons (42 gals = one barrel, 5,000 gals = ~120 bbls).

Rates are expressed in gals/minute, 42 gals/minute = 1 bbl/min, 500 gal/min = ~12 bbls/min.

Flush volumes are based on the total volume of open borehole, therefore as each stage is completed the volume of flush decreases as the volume of borehole is decreased.

Total amount of proppant used is approximately 450,000 pounds

Source: Arthur et al., 2008

Source: Chesapeake Energy Corporation, 2008

Hydraulic Fracturing of a Marcellus Shale Well, West Virginia

Every aspect of the fracture stimulation process is carefully monitored.

Fracturing Fluids and Additives

As described above, the current practice for hydraulic fracture treatments of shale gas reservoirs is to apply a sequenced pumping event in which millions of gallons of water-based fracturing fluids mixed with proppant materials are pumped in a controlled and monitored manner into the target shale formation above fracture pressure[281].

The fracturing fluids used for gas shale stimulations consist primarily of water but also include a variety of additives. The number of chemical additives used in a typical fracture treatment varies depending on the conditions of the specific well being fractured. A typical fracture treatment will use very low concentrations of between 3 and 12 additive chemicals depending on the characteristics of the water and the shale formation being fractured. Each component serves a specific, engineered purpose[282]. The predominant fluids currently being used for fracture treatments in the gas shale plays are water-based fracturing fluids mixed with friction-reducing additives (called slickwater)[283].

The addition of friction reducers allows fracturing fluids and proppant to be pumped to the target zone at a higher rate and reduced pressure than if water alone were used. In addition to friction reducers, other additives include: biocides to prevent microorganism growth and to reduce bio-fouling of the fractures; oxygen scavengers and other stabilizers to prevent corrosion of metal pipes; and acids that are used to remove drilling mud damage within the near-wellbore area[284]. These fluids are used not only to create the fractures in the formation but also to carry a propping agent (typically silica sand) which is deposited in the induced fractures.

Exhibit 35[285] demonstrates the volumetric percentages of additives that were used for a nine-stage hydraulic fracturing treatment of a Fayetteville Shale horizontal well. The make-up of fracturing fluid varies from one geologic basin or formation to another. Evaluating the relative volumes of the components of a fracturing fluid reveals the relatively small volume of additives that are present. The additives depicted on the right side of the pie chart represent less than 0.5% of the total fluid volume. Overall the concentration of additives in most slickwater fracturing fluids is a relatively consistent 0.5% to 2% with water making up 98% to 99.5%.

Because the make-up of each fracturing fluid varies to meet the specific needs of each area, there is no one-size-fits-all formula for the volumes for each additive. In classifying fracturing fluids and their additives it is important to realize that service companies that provide these additives have developed a number of compounds with similar functional properties to be used for the same purpose in different well environments. The difference between additive formulations may be as small as a change in concentration of a specific compound. Although the hydraulic fracturing industry may have a number of compounds that can be used in a hydraulic fracturing fluid, any single fracturing job would only use a few of the available additives. For example, in Exhibit 35 there are 12 additives used, covering the range of possible functions that could be built into a fracturing fluid. It is not uncommon for some fracturing recipes to omit some compound categories if their properties are not required for the specific application.

Most industrial processes use chemicals and almost any chemical can be hazardous in large enough quantities or if not handled properly. Even chemicals that go into our food or drinking water can be hazardous. For example, drinking water treatment plants use large quantities of chlorine. When used and handled properly, it is safe for workers and near-by residents and provides clean, safe drinking water for the community. Although the risk is low, the potential exists for unplanned releases that could have serious effects on human health and the environment. By the same token, hydraulic fracturing uses a number of chemical additives that could be hazardous, but are safe when properly handled according to requirements and long-standing industry practices. In addition, many of these additives are common chemicals which people regularly encounter in everyday life.

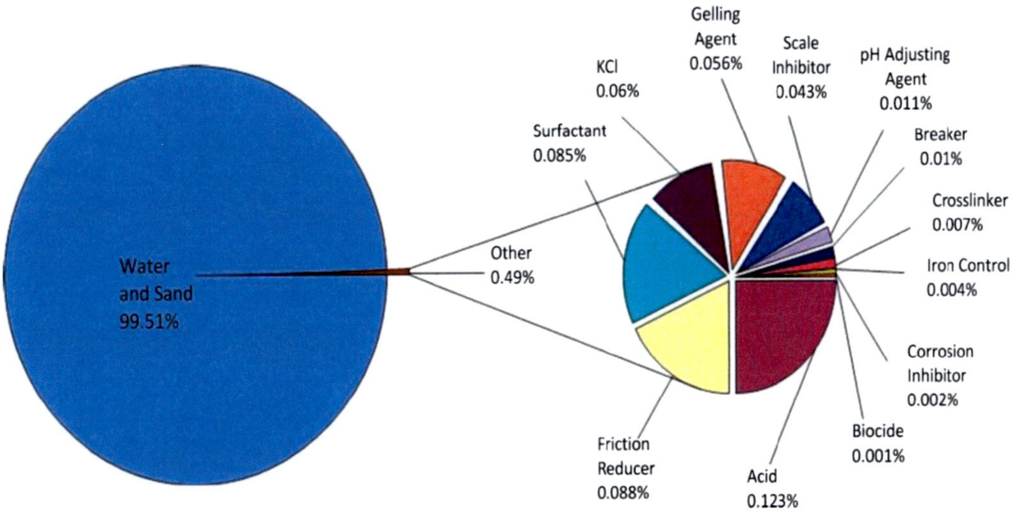

Source: ALL Consulting based on data from a fracture operation in the Fayetteville Shale, 2008

Exhibit 35.Volumetric Composition of a Fracture Fluid

Exhibit 36[286] provides a summary of the additives, their main compounds, the reason the additive is used in a hydraulic fracturing fluid, and some of the other common uses for these compounds. Hydrochloric acid (HCl) is the single largest liquid component used in a fracturing fluid aside from water; while the concentration of the acid may vary, a 15% HCl mix is a typical concentration. A 15% HCl mix is composed of 85% water and 15% acid, therefore, the volume of acid is diluted by 85% with water in its stock solution before it is pumped into the formation during a fracturing treatment. Once the entire stage of fracturing fluid has been injected, the total volume of acid in an example fracturing fluid from the Fayetteville shale was 0.123%, which indicates the fluid had been diluted by a factor of 122 times before it is pumped into the formation. The concentration of this acid will only continue to be diluted as it is further dispersed in additional volumes of water that may be present in the subsurface. Furthermore, if this acid comes into contact with carbonate minerals in the subsurface, it would be neutralized by chemical reaction with the carbonate minerals producing water and carbon dioxide as a byproduct of the reaction.

Exhibit 36. Fracturing Fluid Additives, Main Compounds, and Common Uses

Additive Type	Main Compound(s)	Purpose	Common Use of Main Compound
Diluted Acid (15%)	Hydrochloric acid or muriatic acid	Help dissolve minerals and initiate cracks in the rock	Swimming pool chemical and cleaner
Biocide	Glutaraldehyde	Eliminates bacteria in the water that produce corrosive byproducts	Disinfectant; sterilize medical and dental equipment
Breaker	Ammonium persulfate	Allows a delayed break down of the gel polymer chains	Bleaching agent in detergent and hair cosmetics, manufacture of household plastics
Corrosion Inhibitor	N,n-dimethyl formamide	Prevents the corrosion of the pipe	Used in pharmaceuticals, acrylic fibers, plastics
Crosslinker	Borate salts	Maintains fluid viscosity as temperature increases	Laundry detergents, hand soaps, and cosmetics
Friction Reducer	Polyacrylamide	Minimizes friction between the fluid and the pipe	Water treatment, soil Conditioner
	Mineral oil		Make-up remover, laxatives, and candy
Gel	Guar gum or hydroxyethyl cellulose	Thickens the water in order to suspend the sand	Cosmetics, toothpaste, sauces, baked goods, ice cream
Iron Control	Citric acid	Prevents precipitation of metal oxides	Food additive, flavoring in food and beverages; Lemon Juice ~7% Citric Acid
KCl	Potassium chloride	Creates a brine carrier fluid	Low sodium table salt substitute
Oxygen Scavenger	Ammonium bisulfite	Removes oxygen from the water to protect the pipe from corrosion	Cosmetics, food and beverage processing, water treatment
pH Adjusting Agent	Sodium or potassium carbonate	Maintains the effectiveness of other components, such as crosslinkers	Washing soda, detergents, soap, water softener, glass and ceramics
Proppant	Silica, quartz sand	Allows the fractures to remain open so the gas can escape	Drinking water filtration, play sand, concrete, brick mortar
Scale Inhibitor	Ethylene glycol	Prevents scale deposits in the pipe	Automotive antifreeze, household cleansers, and de-icing agent
Surfactant	Isopropanol	Used to increase the viscosity of the fracture fluid	Glass cleaner, antiperspirant, and hair color

Note: The specific compounds used in a given fracturing operation will vary depending on company preference, source water quality and site-specific characteristics of the target formation. The compounds shown above are representative of the major compounds used in hydraulic fracturing of gas shales.

Water Availability

The drilling and hydraulic fracturing of a horizontal shale gas well may typically require 2to 4 million gallons of water[287], with about 3 million gallons being most common. It should be noted that the volume of water needed may vary substantially between wells. In addition the volume of water needed per foot of wellbore appears to be decreasing as technologies and methods improve over time. Exhibit 37[288] presents a table of estimated per-well water needs for four shale gas plays currently being developed.

Water for drilling and hydraulic fracturing of these wells frequently comes from surface water bodies such as rivers and lakes, but can also come from ground water, private water sources, municipal water, and re-used produced water. Most of the producing shale gas basins occur in areas with moderate to high levels of annual precipitation as shown in Exhibit 38[289]. However, even in areas of high precipitation, due to growing populations, other industrial water demands, and seasonal variation in precipitation, it can be difficult to meet the needs of shale gas development and still satisfy regional needs for water.

While the water volumes needed to drill and stimulate shale gas wells are large, they generally represent a small percentage of the total water resource use in the shale gas basins. Calculations indicate that water use will range from less than 0.1% to 0.8% by basin[290]. This volume is small in terms of the overall surface water budget for an area; however, operators need this water when drilling activity is occurring, requiring that the water be procured over a relatively short period of time. Water withdrawals during periods of low stream flow could affect fish and other aquatic life, fishing and other recreational activities, municipal water supplies, and other industries such as power plants. To put shale gas water use in perspective, the consumptive use of fresh water for electrical generation in the Susquehanna River Basin alone is nearly 150 million gallons per day, while the projected total demand for peak Marcellus Shale activity in the same area is 8.4 million gallons per day[291].

One alternative that states and operators are pursuing is to make use of seasonal changes in river flow to capture water when surface water flows are greatest. Utilizing seasonal flow differences allows planning of withdrawals to avoid potential impacts to municipal drinking water supplies or to aquatic or riparian communities. In the Fayetteville Shale play of Arkansas, one operator is constructing a 500-acre-ft impoundment to store water withdrawals from the Little Red River obtained during periods of high flow (storm events or hydroelectric power generation releases from Greer's Ferry Dam upstream of the intake) when excess water is available[292] (one acre-foot is equivalent to the volume of water required to cover one acre with one foot of water). The project is limited to 1,550 acre-ft of water annually. As additional mitigation, the company has constructed extra pipelines and hydrants to provide portions of this rural area with water for fire protection. Also included is monitoring of in-stream water quality as well as game and non-game fish species in the reach of river surrounding the intake. This design provides a water recovery system similar in concept to what some municipal water facilities use. It will minimize the impact on local water supplies because surface water withdrawals will be limited to times of excess flow in the Little Red River. This project was developed with input from a local chapter of Trout Unlimited, an active conservation organization in the area, and represents an innovative environmental solution that serves both the community and the gas developer.

This project was developed with input from a local chapter of Trout Unlimited, an active conservation organization in the area, and represents an innovative environmental solution that serves both the community and the gas developer.

Exhibit 37. Estimated Water Needs for Drilling and Fracturing Wells in Select Shale Gas Plays

Shale Gas Play	Volume of Drilling Water per well (gal)	Volume of Fracturing Water per well (gal)	Total Volumes of Water per well (gal)
Barnett Shale	400,000	2,300,000	2,700,000
Fayetteville Shale	60,000*	2,900,000	3,060,000
Haynesville Shale	1,000,000	2,700,000	3,700,000
Marcellus Shale	80,000*	3,800,000	3,880,000

* Drilling performed with an/or water-based or oil-based muds for deep horizontal well completions.
Note: These volumes are approximate and may vary substantially between wells.
Source: ALL Consulting from discussions with various operators, 2008

Source: ALL Consulting, 2008

Little Red River, Arkansas

Because the development of shale gas is new in some areas, these water needs may challenge supplies and infrastructure. As operators look to develop new shale gas plays, communication with local water planning agencies can help operators and communities to coexist and effectively manage local water resources. Understanding local water needs can help operators develop a water storage or management plan that will meet with acceptance in neighboring communities. Although the water needed for drilling an individual well may represent a small volume over a large area, the withdrawals may have a cumulative impact to watersheds over the short term. This potential impact can be avoided by working with local water resource managers to develop a plan outlining when and where withdrawals will occur (i.e., avoiding headwaters, tributaries, small surface water bodies, or other sensitive sources).

In some basins, one key to the successful development of shale gas is the identification of water supplies capable of meeting the needs of a development company for drilling and fracturing water without interfering with community needs. While a variety of options exist, the conditions of obtaining water are complex and vary by region and even within a region such that developers will also need to understand local water laws[293] .

One key to the successful development of shale gas is the identification of water supplies capable of meeting the needs of a development company for drilling and fracturing water without interfering with community needs.

Water Management

After a hydraulic fracture treatment, when the pumping pressure has been relieved from the well, the water-based fracturing fluid, mixed with any natural formation water present, begins to flow back through the well casing to the wellhead. This produced water may also contain dissolved constituents from the formation itself. The dissolved constituents are naturally occurring compounds and may vary from one shale play to the next or even by area within a shale play. Initial produced water can vary from fresh (<5,000 ppm Total Dissolved Solids (TDS)) to varying degrees of saline (5,000 ppm to 100,000 ppm TDS[294] or higher). The majority of fracturing fluid is recovered in a matter of several hours to a couple of weeks. In various basins and shale gas plays, the volume of produced water may account for less than 30% to more than 70% of the original fracture fluid volume[295]. In some cases, flow back of fracturing fluid in produced water can continue for several months after gas production has begun[296].

A suite of circumstances explains the disposition of fracturing fluids that are not recovered through production. However, it is important to understand that unrecovered fluids, if any, will remain contained within the target formations. Some of these fluids will occupy macro-porosity (typically natural fracture porosity) in the shale formation and some will occupy the micro-pore space vacated by the gas that is produced. Also, some of the fracturing fluids remain stranded in fractures within the reservoir rock that heal after fracturing, thus preventing the fluids from flowing back to the well. Some of these stranded fluids may flow back to the well in very small volumes over an extended time span. The longer contact time these fluids have with the formation further alters the chemistry of these fluids through increased dissolution of formation minerals, making them similar to the natural formation water. For these reasons it is not possible to unequivocally state that 100% of the fracturing fluids have been recovered or to differentiate flow back water from natural formation water.

Natural formation water has been in contact with the reservoir formation for millions of years and thus contains minerals native to the reservoir rock. The salinity, TDS, and overall quality of formation water vary by geologic basin and specific rock strata. After initial production, produced water can vary from brackish (5,000 ppm to 35,000 ppm TDS), to saline (35,000 ppm to 50,000 ppm TDS), to supersaturated brine (50,000 ppm to >200,000 ppm TDS)[297], and some operators report TDS values greater than 400,000 ppm[298]. The variation in composition changes primarily with changes in the natural formation water chemistry.

States, local governments, and shale gas operators seek to manage produced water in a way that protects surface and ground water resources and, if possible, reduces future demands for fresh water. By pursuing the pollution prevention hierarchy of "Reduce, Re-use, and

Recycle" these groups are examining both traditional and innovative approaches to managing shale gas produced water. This water is currently managed through a variety of mechanisms, including underground injection, treatment and discharge, and recycling. Exhibit 39 summarizes current produced water management practices for the various shale gas basins, and is compiled from data collected from producers and regulatory agencies in these basins.

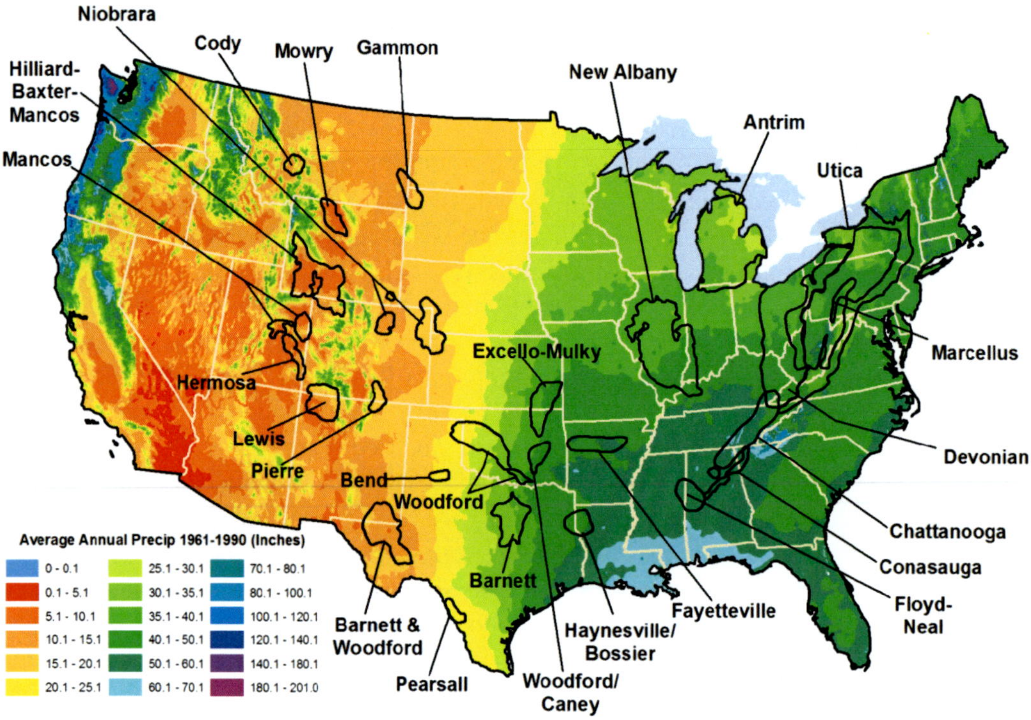

Source: NRCS

Exhibit 38. Annual Rainfall Map of the United States

Underground injection has traditionally been the primary disposal option for oil and gas produced water. In most settings, this may be the best option for shale gas produced water. This process uses salt water disposal wells to place the water thousands of feet underground in porous rock formations that are separated from treatable groundwater by multiple layers of impermeable rock thousands of feet thick. Underground injection of the produced water is not possible in every play as suitable injection zones may not be available. Similar to a producing reservoir, there must be a porous and permeable formation capable of receiving injected fluids nearby. If such is not locally available, it may be possible to transport the produced water to a more distant injection site. In well developed urban plays such as the Barnett Shale around the City of Fort Worth, pipelines have been constructed to transport produced water to injection well disposal sites. This minimizes trucking the water and the resultant traffic, exhaust emissions, and wear on local roads[299]. Injection disposal wells are permitted under the federal Safe Drinking Water Act (SDWA), Underground Injection Control (UIC) program (or in the case of state primacy, under equivalent state programs), a stringently permitted and monitored process with many environmental safeguards in place.

Exhibit 39. Current Produced Water Management by Shale Gas Basin

Shale Gas Basin	Water Management Technology	Availability	Comments
Barnett Shale	Class II injection wells[303]	Commercial and non-commercial	Disposal into the Barnett and underlying Ellenberger Group[304]
	Recycling[305]	On-site treatment and recycling	For reuse in subsequent fracturing jobs [306]
Fayetteville Shale	Class II injection wells[307]	Non-commercial	Water is transported to two injection wells owned and
			operated by a single producing company [308]
	Recycling	On-site recycling	For reuse in subsequent fracturing jobs[309]
Haynesville Shale	Class II injection wells	Commercial and non-commercial	
Marcellus Shale	Class II injection wells	Commercial and non-commercial	Limited use of Class II injection wells[310,311]
	Treatment and discharge	Municipal waste water treatment facilities, commercial facilities reportedly contemplated[312]	Primarily in Pennsylvania
	Recycling	On-site recycling	For reuse in subsequent fracturing jobs[313]
Woodford Shale	Class II injection wells	Commercial	Disposal into multiple confining formations[314]
	Land Application		Permit required through the Oklahoma Corporation Commission[315]
	Recycling	Non-commercial	Water recycling and storage facilities at a central location[316]
Antrim Shale	Class II injection wells	Commercial and non-commercial	
New Albany Shale	Class II injection wells	Commercial and non-commercial	

Treatment of produced water may be feasible through either self-contained systems at well sites or fields or municipal waste water treatment plants or commercial treatment facilities. The availability of municipal or commercial treatment plants may be limited to larger urban areas where treatment facilities with sufficient available capacity already exist. As in underground injection, transportation to treatment facilities may or may not be practical[300].

Re-use of fracturing fluids is being evaluated by service companies and operators to determine the degree of treatment and make-up water necessary for re-use[301]. The practical use of on-site, self-contained treatment facilities and the treatment methods employed will be dictated by flow rate and total water volumes to be treated, constituents and their concentrations requiring removal, treatment objectives and water reuse or discharge requirements. In some cases it would be more practical to treat the water to a quality that could be reused for a subsequent hydraulic fracturing job, or other industrial use, rather than treating to discharge to a surface water body or for use as drinking water. At the time this Primer was developed there were plans to construct commercial waste water treatment facilities specifically designed for the treatment of produced water associated with shale gas development in some locations around the country[302]. The completion and success of such plants no doubt will be closely tied to the successful expansion of production in the various shale gas plays.

New water treatment technologies and new applications of existing technologies are being developed and used to treat shale gas produced water. The treated water can be reused as fracturing make-up water, irrigation water, and in some cases even drinking water. Recycling or re-use of produced water can decrease water demands and provide additional water resources for drought-stricken or arid areas. This allows natural gas-associated produced water to be viewed as a potential resource in its own right[317,318]. In one case, Devon Energy Corporation (Devon) is currently using water distillation units at centralized locations within the Barnett Shale play to treat produced water from hydraulic fracture stimulations[319]. As of early 2008, Devon had hydraulically fractured 50 wells using recycled water. Devon reports that the program is still in its testing and development stages. With further development, such specialized treatment systems may prove beneficial, particularly in more mature plays such as the Barnett; however, their practicality may be limited in emerging shale gas plays. Current levels of interest in recycling and reuse are high, but new approaches and more efficient technologies are needed to make treatment and re-use a wide-spread reality.

While challenges still exist, progress is being made. New technologies and new variations on old technologies are being introduced on a regular basis, and some industry researchers are pursuing ways to reduce the amount of treatment needed. In early 2009, studies were underway to determine the minimum quality of water that could successfully be used in hydraulic fracturing. If hydraulic fracturing procedures or fluid additives can be developed that will allow use of water with a high TDS content, then more treatment options become viable and more water can be re-used. Treatment and re-use of produced water could reduce water withdrawal needs as well as the need for additional disposal options. This approach could also help to resolve many of the concerns associated with these withdrawals.

Naturally Occurring Radioactive Material (NORM)

Some soils and geologic formations contain low levels of radioactive material. This naturally occurring radioactive material (NORM) emits low levels of radiation, to which everyone is exposed on a daily basis. Radiation from natural sources is also called background radiation. Other sources of background radiation include radiation from space and sources that occur naturally in the human body. This background radiation accounts for about 50% of the total exposure for Americans. Most of this background exposure is from radon gas encountered in homes (35% of the total exposure). The average person in the U.S.

is exposed to about 360 millirem (mrem) of radiation from natural sources each year (a mrem, or one one-thousandth of a rem, is a measure of radiation exposure)[320]. The other 50% of exposures for Americans comes primarily from medical sources. Consumer products, industrial, and occupational sources contribute less than 3% of the total exposure[321].

In addition to the background radiation at the earth's surface, NORM can also be brought to the surface in the natural gas production process. When NORM is associated with oil and natural gas production, it begins as small amounts of uranium and thorium within the rock. These elements, along with some of their decay elements, notably $radium_{226}$ and $radium_{228}$[322], can be brought to the surface in drill cuttings and produced water. $Radon_{222}$, a gaseous decay element of radium, can come to the surface along with the shale gas.

When NORM is brought to the surface, it remains in the rock pieces of the drill cuttings, remains in solution with produced water, or, under certain conditions, precipitates out in scales or sludges[323]. The radiation from this NORM is weak and cannot penetrate dense materials such as the steel used in pipes and tanks[324].

The principal concern for NORM in the oil and gas industry is that, over time, it can become concentrated in field production equipment[325] and as sludge or sediment inside tanks and process vessels that have an extended history of contact with formation water[326]. Because the general public does not come into contact with oilfield equipment for extended periods, there is little exposure risk from oilfield NORM. Studies have shown that exposure risks for workers and the public are low for conventional oil and gas operations[327,328].

If measured NORM levels exceed state regulatory levels or OSHA exposure dose risks (29 CFR 1910.1096), the material is taken to licensed facilities for proper disposal. In all cases, OSHA requires employers to evaluate radiation hazards, post caution signs and provide personal protection equipment for workers when radiation doses could exceed 5 mrem in one hour or 100 mrem in any five consecutive days. In addition to these federal worker protections, states have regulations that require operators to protect the safety and health of both workers and the public.

Currently there are no existing federal regulations that specifically address the handling and disposal of NORM wastes[d]. Instead, states producing oil and gas are responsible for promulgating and administering regulations to control the re-use and disposal of NORM-contaminated equipment, produced water, and oil-field wastes. Although regulations vary by state, in general, if NORM concentrations are less than regulatory standards, operators are allowed to dispose of the material by methods approved for standard oilfield waste. Conversely, if NORM concentrations are above regulatory limits, then the material must be disposed of at a licensed facility.

These regulations, standards, and practices ensure that oil and gas operations present negligible risk to the general public with respect to potential NORM exposure. They also present negligible risk to workers when proper controls are implemented[329].

Air Quality

Many of today's air quality rules were primarily designed to regulate emissions from single sources with large volumes of emissions output such as refineries, chemical plants, iron

[d] EPA does have drinking water standards for NORM.

and steel manufacturing facilities, and electrical power generating sites. However, smaller sources such as individual shale gas well sites are also subject to state and federal regulations. Shale gas exploration and production operations are similar to most other conventional and unconventional natural gas exploration and production operations in terms of their air emissions. However, varying gas composition and the fact that there is little or no associated oil production affects the nature of potential emissions.

Sources of Air Emissions

The exploration and production of shale gas may include a variety of potential air emission sources that change depending on the phase of operation. In the early phases of operation, emissions may come from such sources as drilling rigs whose engines may be fueled by either diesel or natural gas and from fracturing operations where multiple diesel-powered pumps are often used to achieve the necessary pressure. Other sources may include the well completion process, which may involve the venting or flaring of some natural gas, and vehicular traffic with engine exhaust and dust from unpaved roads.

Once production has begun, emission sources may include compressors or pumps that may be needed to bring the produced gas up to the surface or up to pipeline pressure. Fugitive emissions such as leaks from pipe connections and associated equipment may also occur. Piping and pumping equipment may include pneumatic instrument systems, which, as part of their normal operations, release or bleed small amounts of natural gas into the atmosphere. Other sources of emissions in this phase of operations include flaring or blow down of gas in non-routine situations, dehydration units to remove water from the produced gas, and sulfur removal systems that may include flares and/or amine units.

Composition of Air Emissions

EPA sets standards, monitors the ambient air across the U.S., and has an active enforcement program to control air emissions from all sources, including the shale gas industry. Although natural gas offers a number of environmental benefits over other sources of energy, particularly other fossil fuels, some air emissions commonly occur during exploration and production activities[330]. These emissions and their sources are discussed below.

As in any construction or industrial activity, NO_x are formed when fossil fuel is burned to provide power to machinery such as compressor engines and during flaring operations. In addition, VOCs may be emitted during the dehydration of natural gas. VOC emissions are typically lower in natural gas activities than those associated with oil production because gas production is essentially a closed process from well to pipeline with fewer opportunities for emissions. In addition, emissions of benzene, toluene, ethylbenzene, and xylenes are low simply because these compounds do not exist in significant quantities in the natural gas stream. The oil and gas industry in general is a lesser contributor to air emissions than numerous other common sources (see Exhibit 40[331]). Further, oil and natural gas production contributes only 2% of the total benzene emissions in the U.S.[332], and shale gas represents a very small subset of this 2%.

Particulate Matter (PM) may occur from dust or soil entering the air during pad construction, traffic on access roads, and diesel exhaust from vehicles and engines. In addition, CO may be emitted during flaring and from the incomplete combustion of carbon-based fuels used in engines. Flaring is seldom necessary with natural gas operations except during short periods of well testing, completions or workovers and non-routine situations such as a temporary pipeline closure. Exhibit 42[333] shows that CO emissions from the natural gas industry represent a very small part of the total[334]

SO_2 may form when fossil fuels containing sulfur are burned. Thus, SO_2 may be emitted from gasoline or diesel powered equipment used at a shale gas production site. However, emissions of SO_2 are typically very small for shale gas operations compared to coal or oil[335].

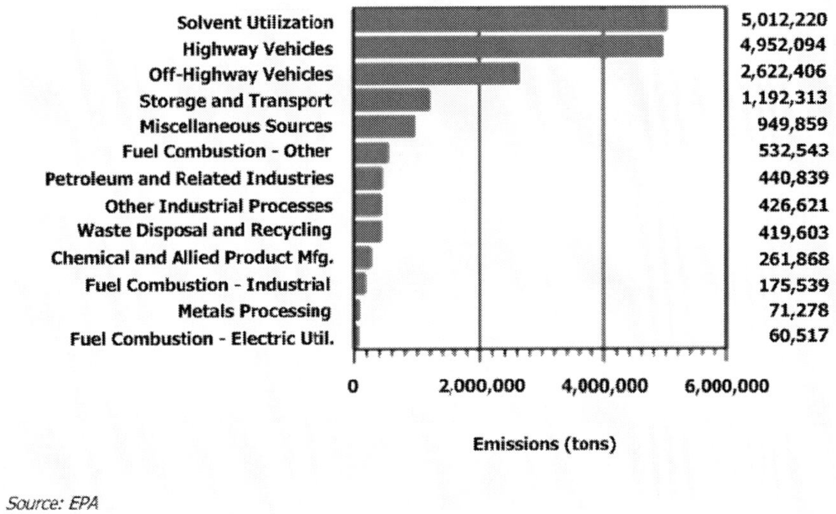

Exhibit 40. VOC Emissions by Source Category

Ozone (O_3) itself is not released directly during natural gas development, but two of its main precursors, volatile organic compounds (VOCs) and NO_x, may combine with sunlight to form ground-level O_3 which can then be associated with exploration and production operations.

Hydrogen sulfide (H_2S) emissions are not a concern in shale gas production as, based on discussions with operators from each of the major basins, the shale gas plays developed to date have not produced "sour" gas. If H_2S is encountered as production continues, both states and operators are well equipped to implement appropriate safety measures. States have well-established public safety and worker protection requirements in place and operators have access to proven procedures for working with natural gas contaminated with H_2S.

The American Petroleum Institute (API) has a Recommended Practice (RP 49) for Drilling and Well Servicing Operations Involving H_2S[336]. Producers voluntarily follow this practice to minimize the release of and exposure to H_2S. In areas where concentrations of H_2S may exceed 10 parts per million (ppm), producers implement an H_2S contingency plan. The plan includes appropriate instruction in the use of hydrogen sulfide safety equipment to all personnel present at all hydrogen sulfide hazard areas, gas detection where hydrogen sulfide may exist, and appropriate respiratory protection for normal and emergency use.

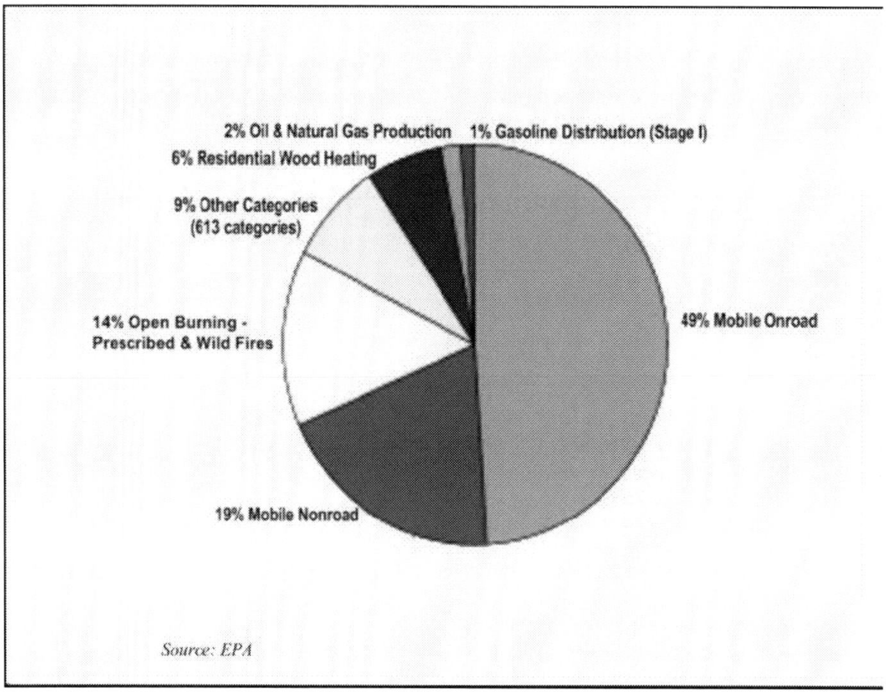

Exhibit 41. Benzene Emissions by Source - 1999

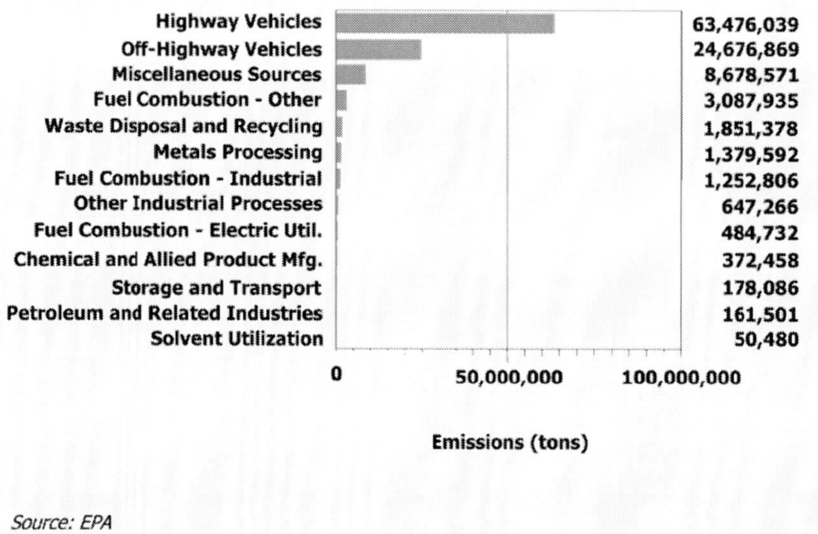

Exhibit 42. Co Emissions by Source Category

Methane (CH_4) is the principal component of natural gas and a known GHG. Although the processing of natural gas is essentially confined from the well to sales, CH_4 may be released as a fugitive emission from gas processing equipment, especially equipment in high pressure service such as pneumatic controls. Producers have strong economic incentives to limit fugitive methane emissions to the greatest degree possible in order to maximize delivery of methane to market. Therefore, they rely on multiple BMPs (e.g., low-bleed gauges and valves, inspection and maintenance programs, infra-red (IR) cameras, etc.[337]) to reduce any potential energy loss.

Another potential source of emissions in natural gas fields are compressor engines. Many gas compressor engines are fueled by natural gas from the lease. Engine manufacturers are constantly improving their technology to reduce the amount of NO_x emissions from their engines. One key has been the use of catalytic technologies to chemically convert NO_x into inert compounds. The addition of catalytic emissions controls has successfully lowered engine emissions from 20 grams per horsepower hour down to 2 grams of NO_x per horsepower hour or less. Also, the addition of air-fuel ratio controllers can be used to ensure the continuous low emissions performance of these engines. Recent EPA regulations require new engines to meet more stringent low NO_x emissions standards regardless of engine size or fuel.

Technological Controls and Practices

The best way to reduce air pollution is to prevent it from occurring in the first place. Pollution prevention can take many forms—upgrading equipment, improving operational practices, reducing waste through byproduct synergies, improving management practices, and installing emissions controls. Several government programs have been established that encompass avoidance, minimization, and mitigation strategies applicable to exploration and production activities. Some are mandatory regulations, as described in the Regulatory Framework section, while others are voluntary.

An example of the latter is the Natural Gas STAR program, a voluntary partnership between the EPA and the natural gas industry formed in 1995 to find cost-effective ways to ensure the natural gas industry is doing everything possible to prevent energy losses and to minimize GHG emissions[338]. The primary goals of the program are to promote technology transfer and implement cost-effective BMPs while reducing CH_4 emissions. The program provides information on many practices that not only reduce CH_4 emissions, but also works to retain greater volumes of natural gas for producers to sell.

Some of the most effective and economic technologies promoted by this program include:

1. Identification of high-bleed pneumatic devices (transducers, valves, controllers, etc.) and replacement with low-bleed ones to reduce fugitive product losses. Traditional pneumatic devices control processes by measuring changes in pressure, releasing small quantities of natural gas in the process. Newer devices are now available that perform the same functions while releasing much smaller amounts of gas.
2. Use of IR cameras in the field to visually identify any fugitive hydrocarbon leaks so that they may be rapidly repaired to reduce potential energy losses. These cameras are tuned to the wavelengths that are reflected by hydrocarbon gases, so that those normally-invisible gases actually become visible as "smoke" in the camera image, thus allowing companies to quickly detect and repair leaks.

3. Installation of flash tank separators in situations that require the use of dehydrators. This can recover 90 to 99% of the methane that would otherwise be flared or vented into the atmosphere[339].
4. Performance of green well completions and workovers. These shale gas operations typically use portable equipment to process and direct the produced natural gas into tanks or directly into the pipeline rather than the traditional practice of venting or flaring the gas. On average, green completions recover 53% of the natural gas that would otherwise have been flared or vented. That captured gas is now retained and sold to market[340].

Many other pollution reduction technologies and practices are described on EPA's GasSTAR web site. In 2004, the Methane to Markets Partnership was formed as a voluntary international program aimed at advancing the recovery and use of methane as a valuable clean energy source[341]. The program includes the oil and gas sector as a focus area along with coal mines, landfills, and the agricultural business. There are approximately 400 program members across the globe representing the oil and gas sector[342]. The collective results of these voluntary programs have been substantial. Total U.S. methane emissions in 2005 were over 11% lower than emissions in 1990, in spite of economic growth over that same time period[343]. EPA expects that these emissions will continue to fall in the future due to expanded industry participation and the ongoing commitment of the participating companies to identify and implement cost-effective technologies and practices.

Additional technologies and practices have been identified that may be used in some settings to reduce air emissions in shale gas fields. One such practice is the use of natural gas instead of diesel to fuel drilling rigs. Another emission-reducing practice applicable to some settings is the use of centralized processing facilities; this reduces vehicle trips, and therefore engine exhaust and dust emissions. Operators have also found that reducing glycol pump rates on dehydration units from their maximum setting to an optimized pump rate will minimize benzene, toluene, ethylbenzene, and total xylenes (BTEX) emissions. These units are often operated at a rate (based on at or near maximum throughput) that accommodates the initial, high rate of gas production from a field. However, as production rates decline, the dehydration units can be adjusted to conform to the lower gas throughput and reduce emissions. Other emission-reducing technologies include the installation of plunger lift systems into shale gas well heads to optimize gas production and reduce methane emissions associated with blowdown operations as well as the optimization of compressor and pump sizes to reduce the necessary horsepower and thus the subsequent exhaust emissions.

As with all operational practices, these BMPs must be applied on a case-by case basis. In some cases a given BMP may actually be counter-productive. In other cases, a particular BMP may create other environmental or operational problems that must be weighed against each other. While each BMP has certain benefits in certain situations, it cannot be universally applied or required.

State and federal requirements along with the technologies and practices developed by industry serve to limit air emissions from shale gas operations. As described earlier, state and federal requirements ensure that local conditions and other emission sources in the area are considered in issuing permits. In addition, advanced technologies and current practices serve to limit air emissions from modern shale gas development.

Summary

The primary differences between modern shale gas development and conventional natural gas development are the extensive use of horizontal drilling and multi-stage hydraulic fracturing. Horizontal drilling allows an area to be developed with substantially fewer wells than would be needed if vertical wells were used. The overall process of horizontal drilling varies little from conventional drilling, with casing and cementing being used to protect fresh and treatable groundwater. The use of horizontal drilling has not introduced new environmental concerns. On the contrary, the reduced number of horizontal wells needed, coupled with multiple wells drilled from a single pad, has significantly reduced surface disturbances and the associated impacts to wildlife and impacts from dust , noise, and traffic. Where shale gas development has intersected with urban and industrial settings, regulators and industry have developed special practices to help reduce community impacts, impacts to sensitive environmental resources, and interference with existing businesses.

Hydraulic fracturing has been a key technology in making shale gas an affordable addition to the Nation's energy supply, and the technology has proven to be a safe and effective stimulation technique. Ground water is protected during the shale gas fracturing process by a combination of the casing and cement that is installed when the well is drilled and the thousands of feet of rock between the fracture zone and any fresh or treatable aquifers. The multi-stage hydraulic fracture operations used in horizontal wells may require 3 to 4 million gallons of water. Since it is a relatively new use in these areas, withdrawals for hydraulic fracturing must be balanced with existing water demands. Once the fracture treatment is completed, most of the fracture water comes back to the surface and must be managed in a way that conserves and protects water resources. While challenges continue to exist with water availability and water management, innovative regional solutions are emerging that allow shale gas development to continue while ensuring that the water needs of other users can be met and that surface and ground water quality is protected.

An additional consideration in shale gas development is the potential for low levels of naturally occurring radioactive material (NORM) to be brought to the surface. While NORM may be encountered in shale gas operations, there is negligible exposure risk for the general public and there are well established regulatory programs that ensure public and worker safety.

Although the use of natural gas offers a number of environmental benefits over other fossil energy sources, some air emissions commonly occur during exploration and production activities. EPA sets standards, monitors the ambient air quality across the U.S., and has an active enforcement program to control air emissions from all sources, including the shale gas industry. Gas field emissions are controlled and minimized through a combination of government regulation and voluntary avoidance, minimization, and mitigation strategies.

Taken together, state and federal requirements, along with the technologies and practices developed by industry, serve to protect human health and to help reduce environmental impacts from shale gas operations.

Acronyms and Abbreviations

API	American Petroleum Institute
Bbls	barrels, petroleum (42 gallons)
Bcf	billion cubic feet
BLM	Bureau of Land Management
BMP	Best Management Practices
Btu	British thermal units
CAA	Clean Air Act
CBNG	Coal Bed Natural Gas
CEQ	Council on Environmental Quality
CFR	Code of Federal Regulations
CERCLA	Comprehensive Environmental Response, Compensation, and Liability Act
CH_4	Methane
CO	Carbon Monoxide
CO_2	Carbon Dioxide
CWA	Clean Water Act
DRBC	Delaware River Basin Commission
EIA	Energy Information Administration
ELG	Effluent Limitation Guidelines
EPA	Environmental Protection Agency
EPCRA	Emergency Planning and Community Right-to-Know Act
FR	Federal Register
Ft	foot/feet
FWS	Fish and Wildlife Service
Gal	gallon
GHG	Greenhouse Gases
GWPC	Ground Water Protection Council
H_2S	Hydrogen Sulfide
HAP	Hazardous Air Pollutant
HCl	Hydrochloric acid
IOGCC	Interstate Oil and Gas Compact Commission
IR	infra-red
Mcf	thousand cubic feet
MMcf	million cubic feet
Mrem	millirem
mrem/yr	millirem per year
MSDSs	Material Safety Data Sheets
NEPA	National Environmental Policy Act
NESHAPs	National Emission Standards for Hazardous Air Pollutants
NETL	National Energy Technology Laboratory
NORM	Naturally Occurring Radioactive Material
NO_x	Nitrogen Oxides
NPDES	National Pollution Discharge Elimination System

NYDEC	New York State Department of Environmental Conservation
O_3	Ozone
OPA	Oil Pollution Act
OSHA	Occupational Safety and Health Administration
PM	Particulate Matter
ppm	parts per million
RAPPS	Reasonable and Prudent Practices for Stabilization
RCRA	Resource Conservation and Recovery Act
RP	Recommended Practice
RQ	Reportable Quantity
SARA	Superfund Amendments and Reauthorization Act
SCF	standard cubic feet
SDWA	Safe Drinking Water Act
SO_2	Sulfur Dioxide
SPCC	Spill Prevention, Control and Countermeasures
SRBC	Susquehanna River Basin Commission
STRONGER	State Review of Oil and Natural Gas Environmental Regulation, Inc.
SWDA	Solid Waste Disposal Act
tcf	trillion cubic feet
TDS	Total Dissolved Solids
tpy	tons per year
TRI	Toxics Release Inventory
UIC	Underground Injection Control
U.S.	United States
U.S.C.	United States Code
USDW	Underground Source of Drinking Water
USGS	United States Geological Survey
VOC	Volatile Organic Compound
WQA	Water Quality Act
yr	year

DEFINITIONS

AIR QUALITY. A measure of the amount of pollutants emitted into the atmosphere and the dispersion potential of an area to dilute those pollutants.

AQUIFER. A body of rock that is sufficiently permeable to conduct groundwater and to yield economically significant quantities of water to wells and springs.

BASIN. A closed geologic structure in which the beds dip toward a central location; the youngest rocks are at the center of a basin and are partly or completely ringed by progressively older rocks.

BIOGENIC GAS. Natural gas produced by living organisms or biological processes.

CASING. Steel piping positioned in a wellbore and cemented in place to prevent the soil or rock from caving in. It also serves to isolate fluids, such as water, gas, and oil, from the surrounding geologic formations.

COAL BED METHANE/NATURAL GAS (CBM/CBNG). A clean-burning natural gas found deep inside and around coal seams. The gas has an affinity to coal and is held in place by pressure from groundwater. CBNG is produced by drilling a wellbore into the coal seam(s), pumping out large volumes of groundwater to reduce the hydrostatic pressure, allowing the gas to dissociate from the coal and flow to the surface.

COMPLETION. The activities and methods to prepare a well for production and following drilling. Includes installation of equipment for production from a gas well.

CORRIDOR. A strip of land through which one or more existing or potential utilities may be co-located.

DISPOSAL WELL. A well which injects produced water into an underground formation for disposal.

DIRECTIONAL DRILLING. The technique of drilling at an angle from a surface location to reach a target formation not located directly underneath the well pad.

DRILL RIG. The mast, draw works, and attendant surface equipment of a drilling or workover unit.

EMISSION. Air pollution discharge into the atmosphere, usually specified by mass per unit time.

ENDANGERED SPECIES. Those species of plants or animals classified by the Secretary of the Interior or the Secretary of Commerce as endangered pursuant to Section 4 of the Endangered Species Act of 1973, as amended. See also **Threatened and Endangered Species.**

EXPLORATION. The process of identifying a potential subsurface geologic target formation and the active drilling of a borehole designed to assess the natural gas or oil.

FLOW LINE. A small diameter pipeline that generally connects a well to the initial processing facility.

FORMATION (GEOLOGIC). A rock body distinguishable from other rock bodies and useful for mapping or description. Formations may be combined into groups or subdivided into members.

FRACTURING FLUIDS. A mixture of water and additives used to hydraulically induce cracks in the target formation.

GROUND WATER. Subsurface water that is in the zone of saturation; source of water for wells, seepage, and springs. The top surface of the groundwater is the "water table."

HABITAT. The area in which a particular species lives. In wildlife management, the major elements of a habitat are considered to be food, water, cover, breeding space, and living space.

HORIZONTAL DRILLING. A drilling procedure in which the wellbore is drilled vertically to a kick-off depth above the target formation and then angled through a wide 90 degree arc such that the producing portion of the well extends horizontally through the target formation.

HYDRAULIC FRACTURING. Injecting fracturing fluids into the target formation at a force exceeding the parting pressure of the rock thus inducing a network of fractures through which oil or natural gas can flow to the wellbore.

HYDROSTATIC PRESSURE. The pressure exerted by a fluid at rest due to its inherent physical properties and the amount of pressure being exerted on it from outside forces.

INJECTION WELL. A well used to inject fluids into an underground formation either for enhanced recovery or disposal.

LEASE. A legal document that conveys to an operator the right to drill for oil and gas. Also, the tract of land, on which a lease has been obtained, where producing wells and production equipment are located.

NORM (Naturally Occurring Radioactive Material). Low-level, radioactive material that naturally exists in native materials.

ORIGINAL GAS- IN- PLACE The entire volume of gas contained in the reservoir, regardless of the ability to produce it

PARTICULATE MATTER (PM). A small particle of solid or liquid matter (e.g., soot, dust, and mist). PMio refers to particulate matter having a size diameter of less than 10 millionths of a meter (micrometer) and PM2.5 being less than 2.5 micro-meters in diameter.

PERMEABILITY. A rock's capacity to transmit a fluid; dependent upon the size and shape of pores and interconnecting pore throats. A rock may have significant porosity (many microscopic pores) but have low permeability if the pores are not interconnected. Permeability may also exist or be enhanced through fractures that connect the pores.

PRIMACY. A right that can be granted to state by the federal government that allows state agencies to implement programs with federal oversight Usually, the states develop their own set of regulations. By statute, states may adopt their own standards, however, these must be at least as protective as the federal standards they replace, and may be even more protective in order to address local conditions. Once these state programs are approved by the relevant federal agency (usually the EPA), the state then has primacy jurisdiction.

PRODUCED WATER. Water produced from oil and gas wells.

PROPPING AGENTS/PROPPANT. Silica sand or other particles pumped into a formation during a hydraulic fracturing operation to keep fractures open and maintain permeability.

PROVED RESERVES That portion of recoverable resources that is demonstrated by actual production or conclusive formation tests to be technically, economically, and legally producible under existing economic and operating conditions.

RECLAMATION. Rehabilitation of a disturbed area to make it acceptable for designated uses. This normally involves regrading, replacement of topsoil, re-vegetation, and other work necessary to restore it.

SET-BACK. The distance that must be maintained between a well or other specified equipment and any protected structure or feature.

SHALE GAS. Natural gas produced from low permeability shale formations.

SLICKWATER. A water based fluid mixed with friction reducing agents, commonly potassium chloride.

SOLID WASTE. Any solid, semi-solid, liquid, or contained gaseous material that is intended for disposal.

SPLIT ESTATE. Condition that exists when the surface rights and mineral rights of a given area are owned by different persons or entities; also referred to as "severed estate".

STIMULATION. Any of several processes used to enhance near wellbore permeability and reservoir permeability.

STIPULATION. A condition or requirement attached to a lease or contract, usually dealing with protection of the environment, or recovery of a mineral.

SULFUR DIOXIDE (SO_2). A colorless gas formed when sulfur oxidizes, often as a result of burning trace amounts of sulfur in fossil fuels.

TECHNICALLY RECOVERABLE RESOURCES The total amount of resource, discovered and undiscovered, that is thought to be recoverable with available technology, regardless of economics.

THERMOGENIC GAS. Natural gas that is formed by the combined forces of high pressure and temperature (both from deep burial within the earth's crust), resulting in the natural cracking of the organic matter in the source rock matrix.

THREATENED AND ENDANGERED SPECIES. Plant or animal species that have been designated as being in danger of extinction. See also **Endangered Species.**

TIGHT GAS. Natural gas trapped in a hardrock, sandstone or limestone formation that is relatively impermeable.

TOTAL DISSOLVED SOLIDS (TDS). The dry weight of dissolved material, organic and inorganic, contained in water and usually expressed in parts per million.

UNDERGROUND INJECTION CONTROL PROGRAM (UIC). A program administered by the Environmental Protection Agency, primacy state, or Indian tribe under the Safe Drinking Water Act to ensure that subsurface emplacement of fluids does not endanger underground sources of drinking water.

UNDERGROUND SOURCE OF DRINKING WATER (USDW). 40 CFR Section 144.3 An aquifer or its portion:

(a) (1) Which supplies any public water system; or

(2) Which contains a sufficient quantity of ground water to supply a public water system; and

(i) Currently supplies drinking water for human consumption; or

(ii) Contains fewer than 10,000 mg/1 total dissolved solids; and

(b) Which is not an exempted aquifer.

WATER QUALITY. The chemical, physical, and biological characteristics of water with respect to its suitability for a particular use.

WATERSHED. All lands which are enclosed by a continuous hydrologic drainage divide and lay upslope from a specified point on a stream.

WELL COMPLETION. See **Completion.**

WORKOVER. To perform one or more remedial operations on a producing or injection well to increase production. Deepening, plugging back, pulling, and resetting the liner are examples of workover operations.

End Notes

[1] EIA. 2008. *Annual Energy Outlook 2008 with Projections to 2030* http://www.eia.doe.gov/oiaf/aeo/pdf/0383(2008).pdf. June 2008.

[2] Ibid.

[3] EIA. 2007. Annual Energy Review 2007, Tables 4.1 and 4.2. http://www.eia.doe.gov/emeu/aer/resource.html. June 2008.

[4] EIA. 2009. *U.S. Crude Oil, Natural Gas, and Natural Gas Liquids Reserves 2007 Annual Report.* http://www.eia.doe.gov/oil_gas/natural_gas/data_publications/crude_oil_natural_gas_reserves/cr.html. February 2009.

[5] Navigant Consulting. 2008. *North American Natural Gas Supply Assessment.* Prepared for American Clean Skies Foundation. July 4, 2008.

[6] EIA. 2007. Annual Energy Review 2007, Table 6.1. http://www.eia.doe.gov/emeu/aer/natgas.html. June 2008.

[7] EIA. 2009. *March 2009 Monthly Energy Review, Table 4.3.* http://www.eia.doe.gov/emeu/mer/natgas.html. March 2009.

[8] EIA. 2008. *Annual Energy Outlook 2008 with Projections to 2030* http://www.eia.doe.gov/oiaf/aeo/pdf/0383(2008).pdf. June2008.

[9] EIA. 1999. *Natural Gas 1998: Issues and Trends* http://www.eia.doe.gov/pub/oil_gas/natural_gas/analysis_publications/natural_gas_1998_issues_trends/pdf/it98.pdf. April 1999.

[10] NaturalGas.org. *Electrical Generation Using Natural Gas.* http://www.naturalgas.org/overview/uses_eletrical.asp. Accessed: September 2008.

[11] EIA. 2007. Annual Energy Review 2007, Table 8.11b. http://www.eia.doe.gov/emeu/aer/elect.html. June 2008.

[12] EIA. 2002 Manufacturing Energy Consumption Survey. http://www.eia.doe.gov/emeu/mecs/mecs2002/data02/shelltables.html. March 2005.

[13] U.S. Bureau of Labor Statistics. 2007. Monthly Labor Review, November 2007. http://www.bls.gov/opub/mlr/2007/11/contents.htm.

[14] EIA. *Natural Gas Basic Statistics.* http://www.eia.doe.gov/basics/quickgas.html. Accessed: September 2008.

[15] Independent Petroleum Association of Mountain States (IPAMS). America's Independent Natural Gas Producers. *Producing Today's Clean Energy, Ensuring Tomorrow's Innovation.* http://www.ipams.org/media/docs/Callupdraft10.pdf. Accessed: September 2008.

[16] Ambrose, W. A., E. C. Potter, & R. Briceno. 2008. *An Unconventional Future for Natural Gas in the United States.* http://www.geotimes.org/feb08/article.html?id=feature_gas.html. February 2008.

[17] EIA. 2008 *Natural Gas Summary.* http://tonto.eia.doe.gov/dnav/ng/ng_sum_lsum_dcu_nus_a.htm. Accessed June 2008. .

[18] EIA. 2008. *Annual Energy Outlook 2008 with Projections to 2030.* http://www.eia.doe.gov/oiaf/aeo/pdf/0383(2008).pdf. June 2008.

[19] EIA. 2007. *International Energy Outlook 2007, Chapter 4: Natural Gas.* http://tonto.eia.doe.gov/FTPROOT/forecasting/0484(2007).pdf. May 2007.

[20] EIA. 1999. *Natural Gas 1998: Issues and Trends.* http://www.eia.doe.gov/pub/oil_gas/natural_gas/analysis_publications/natural_gas_1998_issues_trends/pdf/it98.pdf. April 1999.

[21] Ibid.

[22] NaturalGas.org. 2008. *Natural Gas and the Environment.* http://www.naturalgas.org/environment/naturalgas.asp. Accessed: September 2008.

[23] American Clean Skies Foundation. *Natural Gas Myth vs. Fact.* www.CleanSkies.org. Accessed: September 2008.

[24] EIA. 2008. *Greenhouse Gases, Climate Change, and Energy.* http://www.eia.doe.gov/bookshelf/brochures/greenhouse/Chapter1.htm. May 2008.

[25] EIA. 2008. *Annual Energy Outlook 2008 with Projections to 2030.* http://www.eia.doe.gov/oiaf/aeo/pdf/0383(2008).pdf. June 2008.

[26] Independent Petroleum Association of Mountain States (IPAMS). America's Independent Natural Gas Producers. *Producing Today's Clean Energy, Ensuring Tomorrow's Innovation.* http://www.ipams.org/media/docs/Callupdraft10.pdf. Accessed: September 2008.

[27] Public Lands Advocacy. 2008. *Energy Insecurity.* Presentation at Public Lands Advocacy NEPA-Permitting Seminar/2008 Annual Meeting. http://www.publiclandsadvocacy.org/presentations.htm. June 11, 2008.

[28] American Clean Skies Foundation. *U.S. Fuel Goals.* www.cleanskies.org. Accessed: September 2008.

[29] Chemistry and Technology of Fuels and Oils. 2000. Volume 36, Number 2, pp 82-88. March 2000.

[30] NaturalGas.org. *Overview of Natural Gas.* Background. www.naturalgas.org/overview/background.asp. Accessed: September 2008.

[31] Ibid.

[32] Ibid.

[33] EIA. *About U.S. Natural Gas Pipelines – Transporting Natural Gas.* www.eia.doe.gov/pub/oil_gas/natural_gas/analysis_publications/ngpipeline. Accessed: September 2008.

[34] EIA. *Glossary.* www.eia.doe.gov/glossary/glossary_b.htm. Accessed: September 2008.

[35] NaturalGas.org. *Overview of Natural Gas.* Background. www.naturalgas.org/overview/background.asp. Accessed: September 2008.

[36] Oklahoma Natural Gas Company. 2005. Rate Schedule 701. *Compressed Natural Gas Service. Individual Metering.* http://www.oneok.com/ong/pdf/rate_schedule/701.pdf. October 4, 2005.

[37] NW Natural. *Science Facts.* www.nwnatural.com/content_safety.asp?id=297. Accessed: September 2008.

[38] BP. 2008. *Statistical Review of World Energy – 2008. Natural Gas.* http://www.bp.com/subsection.do?categoryId=9023762&contentId=7044550. Accessed: September 2008.

[39] Navigant Consulting. 2008. *North American Natural Gas Supply Assessment.* Prepared for American Clean Skies Foundation. July 4, 2008.

[40] EIA. 2009. *U.S. Crude Oil, Natural Gas, and Natural Gas Liquids Reserves 2007 Annual Report*. Available online: http://www.eia.doe.gov/oil_gas/natural_gas/data_publications/crude_oil_natural_gas_reserves/cr.html. Accessed February 2009.
[41] Williams, P. 2008. American Clean Skies. *A Vast Ocean of Natural Gas*. p 44-50. Summer 2008.
[42] EIA. 2008. *Annual Energy Outlook 2008 with Projections to 2030*. http://www.eia.doe.gov/oiaf/aeo/pdf/0383(2008).pdf.June 2008.
[43] Navigant Consulting. 2008. *North American Natural Gas Supply Assessment*. Prepared for American Clean Skies Foundation. July 4, 2008.
[44] ALL Consulting modified from USGS and other sources. March 2009.
[45] EIA. 2008. *Is U.S. Natural Gas Production Increasing*? http://tonto.eia.doe.gov/energy_in_brief/natural_gas_production.cfm. June 11, 2008.
[46] American Clean Skies Foundation. 2008. *Natural Gas Planet*. Summer 2008.
[47] EIA. 2008. *Is U.S. Natural Gas Production Increasing*? http://tonto.eia.doe.gov/energy_in_brief/natural_gas_production.cfm.June 11, 2008
[48] Navigant Consulting. 2008. *North American Natural Gas Supply Assessment*. Prepared for American Clean Skies Foundation. July 4, 2008.
[49] Ibid.
[50] Ibid.
[51] Ibid.
[52] Ibid.
[53] CERA. The Shale Gale: The Implications for North American Natural Gas Pipeline Development. March 2009.
[54] Independent Petroleum Association of Mountain States (IPAMS). America's Independent Natural Gas Producers. *Producing Today' s Clean Energy, Ensuring Tomorrow' s Innovation*. http://www.ipams.org/media/docs/Callupdraft10.pdf. Accessed:September 2008.
[55] Harper, J. 2008. *The Marcellus Shale – An Old "New" Gas Reservoir in Pennsylvania*. v 28, no 1. Published by the Bureau of Topographic and Geologic Survey, Pennsylvania Department of Conservation and Natural Resources. Pennsylvania Geology. Spring 2008.
[56] Schlumberger. 2005. *Shale Gas White Paper*. 05-OF299. Modified from: J. K., Frantz and V. Jochen. Schlumberger Marketing Communications. October 2005.
[57] NY DEC. 1992. Division of Mineral Resources. *Final Generic Environmental Impact Statement on the Oil, Gas and Solution Mining Regulatory Program*. 1992.
[58] Harper, J. 2008. *The Marcellus Shale – An Old "New" Gas Reservoir in Pennsylvania*. v 28, no 1. Published by the Bureau of Topographic and Geologic Survey, Pennsylvania Department of Conservation and Natural Resources. Pennsylvania Geology. Spring 2008.
[59] Ibid.
[60] Engelder, T., and G. G. Lash. 2008. *Marcellus Shale Play' s Vast Resource Potential Creating Stir in Appalachia*. The American Oil & Gas Reporter. May 2008.
[61] Williams, P. 2005. *Big Sandy: Kentucky' s Big Sandy Field Remains a Locus of Exploration Activity, More than 90 years after its Discovery*. Oil and Gas Investor. August 2005.
[62] Harrison, W. 2006. *Production History and Reservoir Characteristics of the Antrim Shale Gas Play, Michigan Basin*. Western Michigan University. 2006.
[63] Hayden, J., and D. Pursell, D. 2005. Pickering Energy Partners Inc. *The Barnett Shale. Visitor' s Guide to the Hottest Gas Play in the US*. October 2005.
[64] Ibid.
[65] Halliburton Energy Services. 2008. *U.S. Shale Gas: An Unconventional Resource*. Unconventional Challenges. 2008.
[66] Hayden, J., and D. Pursell, D. 2005. Pickering Energy Partners Inc. *The Barnett Shale. Visitor' s Guide to the Hottest Gas Play in the US*. October 2005.
[67] Cohen, D. 2008. Energy Bulletin. *An Unconventional Play in the Bakken*. April 2008.
[68] USGS. *Assessment of Undiscovered Oil Resources in the Devonian-Mississippian Bakken Formation, Williston Basin Province, Montana and North Dakota*. USGS fact sheet 2008-3021.
[69] Harper, J. 2008. *The Marcellus Shale – An Old "New" Gas Reservoir in Pennsylvania*. v 28, no 1. Published by the Bureau of Topographic and Geologic Survey, Pennsylvania Department of Conservation and Natural Resources. Pennsylvania Geology. Spring 2008.
[70] Ameri, S., K. Aminian, J. A. Miller, D. Doricich, and A. B. Yost. *A Systematic Approach for Economic Development of the Devonian Shale Gas Resources*. SPE 14504.
[71] Halliburton Energy Services. 2008. *U.S. Shale Gas: An Unconventional Resource*. Unconventional Challenges. 2008.
[72] Airhart, M. Geology.com. *The Barnett Shale Gas Boom: Igniting a Hunt for Unconventional Natural Gas Resources*. http://geology.com/research/barnett-shale-gas.shtml. Accessed: September 2008.

[73] Boyer, C., J. Kieschnick, R. Suarez-Rivera, R. Lewis, and G. Walter. 2006. Schlumberger. Oilfield Review. *Producing Gas from Its Source*. Autumn 2006.
[74] Schlumberger. 2005. *Shale Gas White Paper*. 05-OF299. Modified from: J. K., Frantz and V. Jochen. Schlumberger Marketing Communications. October 2005.
[75] Davis ,R., Jr. 1992. *Depositional Systems: An Introduction to Sedimentology and Stratigraphy*. Prentice Hall. 2nd Edition. 1992.
[76] Freeze, R., and J. Cherry. 1979. *Groundwater*. Prentice Hall. 1979.
[77] Nuttal, B.C. 2007. *Kentucky Geological Survey. Predicting Cumulative Production of Devonian Shale Gas Wells from Early Well Performance Data,* Appalachian Basin of Eastern Kentucky. September 2007.
[78] NaturalGas.org. *Unconventional Natural Gas Resources*. Accessed: September 2008.
[79] ALL Consulting and the Montana Board of Oil and Gas Conservation. 2004. *Coal Bed Methane Primer: New Source of Natural Gas – Environmental Implications*. Prepared for the U.S. Department of Energy, National Petroleum Technology Office. February 2004.
[80] Frantz, J.K., and Jochen, V. 2005. Schlumberger. *Shale Gas White Paper*. 05-OF299. Schlumberger Marketing Communications. October 2005.
[81] Geology.com. *Marcellus Shale- Appalachian Basin Natural Gas Play: New Research Results Surprise Everyone on the Potential of this Well-known Devonian Black Shale*. http://geology.com/articles/marcellus-shale.shtml. Accessed: September 2008.
[82] Hayden, J., and D. Pursell. 2005. Pickering Energy Partners, Inc. *The Barnett Shale: Visitor ' s Guide to the Hottest Gas Play in the US*. October 2005.
[83] Halliburton Energy Services. 2008. *U.S. Shale Gas: An Unconventional Resource*. Unconventional Challenges. 2008.
[84] Ibid.
[85] Ibid.
[86] Navigant Consulting. 2008. *North American Natural Gas Supply Assessment*. Prepared for American Clean Skies Foundation. July 4, 2008.
[87] Hayden, J., and D. Pursell. 2005. Pickering Energy Partners, Inc. *The Barnett Shale: Visitor ' s Guide to the Hottest Gas Play in the US*. October 2005.
[88] Ibid.
[89] Ibid.
[90] Ibid.
[91] Boughal, K. 2008. *Unconventional Plays Grow in Number After Barnett Shale Blazed the Way*. World Oil Magazine. v 229, no 8. August 2008.
[92] Berman, A. 2008. *The Haynesville Shale Sizzles with the Barnett Cools*. World Oil Magazine. v 229, no 9. September 2008.
[93] Drilling Contractor. 2000. *Alabama Lawsuit Poses Threat to Hydraulic Fracturing Across U.S.* pp 42-43. January/February 2000.
[94] Haines, L. 2006. Supplement to Oil & Gas Investor. *Shale Gas: Activity Builds in the Woodford Shale*. p 17. http://www.oilandgasinvestor.com/pdf/ShaleGas.pdf. January 2006.
[95] Hayden, J., and D. Pursell. 2005. Pickering Energy Partners, Inc. *The Barnett Shale: Visitor ' s Guide to the Hottest Gas Play in the US*. October 2005.
[96] Ibid.
[97] Arkansas Oil and Gas Commission. 2008. Field Rules and Rule B-15.
[98] Hayden, J., and D. Pursell. 2005 . Pickering Energy Partners, Inc. *The Barnett Shale: Visitor ' s Guide to the Hottest Gas Play in the US*. October 2005.
[99] Ibid.
[100] Berman, A. 2008. *The Haynesville Shale Sizzles while the Barnett Cools*. World Oil Magazine. v 229, no 9. September 2008.
[101] Nyahay, R., J. Leone, L. Smith, J. Martin, and D. Jarvie. 2007. *Update on Regional Assessment of Gas Potential in the Devonian Marcellus and Ordovician Utica Shales of New York*. http://www.searchanddiscovery.net/documents/2007/07101nyahay/index.htm. October 1, 2007.
[102] Cardott, B. 2004. Oklahoma Geological Survey. *Overview of Unconventional Energy Resources of Oklahoma*. http://www.ogs.ou.edu/fossilfuels/coalpdfs/UnconventionalPresentation.pdf. March 9, 2004.
[103] Hayden, J., and D. Pursell. 2005. Pickering Energy Partners ,Inc. *The Barnett Shale: Visitor ' s Guide to the Hottest Gas Play in the US*. October 2005.
[104] Ibid.
[105] Ibid.
[106] Ibid.
[107] Berman, A. 2008. *The Haynesville Shale Sizzles while the Barnett Cools*. World Oil Magazine. v 229, no 9. September 2008.

[108] Soeder, D.J. 1986. *Porosity and Permeability of Eastern Devonian Gas Shale*. SPE Formation Evaluation. v 3, no 1. January 1, 1986.
[109] Vulgamore, T., T. Clawson, C. Pope, S. Wolhart, M. Mayerhofer, S. Machovoe, and C. Waltman. 2007. *Applying Hydraulic Fracture Diagnostics to Optimize Stimulations in the Woodford Shale*. SPE 110029. 2007.
[110] Hayden, J., and D. Pursell. 2005. Pickering Energy Partners ,Inc. *The Barnett Shale: Visitor' s Guide to the Hottest Gas Play in the US.* October 2005.
[111] Ibid.
[112] Ibid.
[113] Ibid.
[114] Petroleum Listing Services. 2008. *Other Players Reporting Haynesville Success*. August 15, 2008.
[115] Soeder, D.J. 1986. *Porosity and Permeability of Eastern Devonian Gas Shale*. SPE Formation Evaluation. v 3, no 1. January 1, 1986.
[116] Jochen, V. Schlumberger. 2006. *New Technology Needs to Produce Unconventional Gas*. November 29, 2006.
[117] Hayden, J., and D. Pursell. 2005. Pickering Energy Partners ,Inc. *The Barnett Shale: Visitor' s Guide to the Hottest Gas Play in the US.* October 2005.
[118] Ibid.
[119] Ibid.
[120] Ibid.
[121] Ibid.
[122] Sumi, L. 2008. Oil and Gas Accountability Project (OGAP). *Shale Gas: Focus on the Marcellus Shale*. http://www.earthworksaction.org/pubs/OGAPMarcellusShaleReport-6-12-08.pdf. May 2008.
[123] Ibid.
[124] Ibid.
[125] Hayden, J., and D. Pursell. 2005. Pickering Energy Partners ,Inc. *The Barnett Shale: Visitor' s Guide to the Hottest Gas Play in the US.* October 2005.
[126] Ibid.
[127] Navigant Consulting. 2008. *North American Natural Gas Supply Assessment*. Prepared for American Clean Skies Foundation. July 4, 2008.
[128] Ibid.
[129] Hayden, J., and D. Pursell. 2005. Pickering Energy Partners ,Inc. *The Barnett Shale: Visitor' s Guide to the Hottest Gas Play in the US.* October 2005.
[130] AAPG. 1987. *Correlation of Stratigraphic Units of North America (CASUNA) Project*. Texas-Oklahoma Tectonic Region. CSD No. 420, Col 9. 1987.
[131] Modified from: Frantz, J.K., and V. Jochen. 2005. Schlumberger. *Shale Gas White Paper*. 05-OF299. Schlumberger Marketing Communications. October 2005.
[132] Railroad Commission of Texas. 2009. *Newark, East (Barnett Shale) Field.* http://www.rrc.state.tx.us/data/fielddata/barnettshale.pdf. Accessed March 2009.
[133] Parshall, J. 2008. *Barnett Shale Showcases Tight-gas Development*. Journal of Petroleum Technology. September 19, 2008.
[134] Halliburton Energy Services. 2008. *U.S. Shale Gas: An Unconventional Resource*. Unconventional Challenges. 2008.
[135] Modified from: Frantz, J.K., and V. Jochen. 2005. Schlumberger. *Shale Gas White Paper*. 05-OF299. Schlumberger Marketing Communications. October 2005.
[136] University of Arkansas. 2008. *Projecting the Economic Impact of the Fayetteville Shale Play for 2008-2012.* Sam M. Walton College of Business. March 2008.
[137] Boughal, K. 2008. *Unconventional Plays Grow in Number After Barnett Shale Blazed the Way*. World Oil Magazine. v 229, no 8. August 2008.
[138] Ibid.
[139] Williams, P. 2008 . American Clean Skies. *A Vast Ocean of Natural Gas*. p 44-50. Summer 2008.
[140] Hillwood International Energy. 2007. *Fayetteville Shale Power.* Arkansas, USA. http://hillwoodenergy.com/Portals/0/Content/Fayetteville_Shale.pdf. July 12, 2007.
[141] Modified from: Frantz, J.K., and V. Jochen. 2005. Schlumberger. *Shale Gas White Paper*. 05-OF299. Schlumberger Marketing Communications. October 2005.
[142] Durham, L. 2008. *Louisiana Play a „Company Maker?* AAPG Explorer. July 2008.
[143] Johnston, J., III, P. Heinrich, J. Lovelace, R. McCulloh, and R. Zimmerman. Louisiana Geological Survey. *Stratigraphic Charts of Louisiana*. Folio series no 8. 2000.
[144] Modified from: USGS. Energy Resources Program. *National Oil and Gas Assessment.* http://certmapper.cr.usgs.gov/noga/broker.jsp?theServlet=NogaNewGISResultsServ&theProvince=67&thePage=play. Accessed: September 2008.
[145] West Virginia Geological and Economic Survey. 1997. *Enhancement of the Appalachian Basin Devonian Shale Resource Base in the GRI Hydrocarbon Model*. Prepared for: Gas Research Institute. December 1997.

[146] Harper, J. 2008. Published by the Bureau of Topographic and Geologic Survey, Pennsylvania Department of Conservation and Natural Resources. Pennsylvania Geology. *The Marcellus Shale – An Old "New" Gas Reservoir in Pennsylvania*. v 28, no 1. Spring 2008.

[147] Ibid.

[148] Arthur, J.D., B. Bohm, and M. Layne. 2008. ALL Consulting. *Hydraulic Fracturing Considerations for Natural Gas Wells of the Marcellus Shale*. Presented at the GWPC Annual Forum in Cincinnati, OH. September 2008.

[149] Modified from: Frantz, J.K., and V. Jochen. 2005. Schlumberger. *Shale Gas White Paper*. 05-OF299. Schlumberger Marketing Communications. October 2005.

[150] Cardott, B. 2007. Oklahoma Geological Survey. *Overview of Woodford Gas-Shale Play in Oklahoma*. 2007.

[151] Boughal, K. 2008. *Unconventional Plays Grow in Number After Barnett Shale Blazed the Way*. World Oil Magazine. v 229, no 8. August 2008.

[152] Cardott, B. 2007. Oklahoma Geological Survey. *Overview of Woodford Gas-Shale Play in Oklahoma*. 2007.

[153] AAPG. 1983. Correlation of Stratigraphic Units of North America (CASUNA) Project. Southwest/Southwest Mid-Continent Region. CSD No. 360, Col 27. 1983.

[154] Modified from: Frantz, J.K., and V. Jochen. 2005. Schlumberger. *Shale Gas White Paper*. 05-OF299. Schlumberger Marketing Communications. October 2005.

[155] Harrison, W. 2006. *Production History and Reservoir Characteristics of the Antrim Shale Gas Play, Michigan Basin*. Western Michigan University. 2006.

[156] Hayden, J., and D. Pursell. 2005. Pickering Energy Partners, Inc. *The Barnett Shale: Visitor' s Guide to the Hottest Gas Play in the US.* October 2005.

[157] Catacosinos, P., W. Harrison III, R. Reynolds, D. Westjohn, and M. Wollensak. 2000. Michigan Department of Environmental Quality (DEQ), Geological Survey Division, and Michigan Basin Geological Survey. *Stratigraphic Nomenclature for Michigan*. 2000.

[158] Indiana Geological Survey. Indiana University. *General Stratigraphic Column for Paleozoic Rocks in Indiana*. Adapted from Shaver, H., et al. 1986. Indiana Geological Survey Bulletin 59.

[159] Illinois State Geological Survey. GeoNote 1. *Mississipian Rocks in Illinois*. http://www.isgs.uiuc.edu/maps-datapub/publications/geonotes/pdf-files/GeoNote1.pdf.

[160] Modified from: Frantz, J.K., and V. Jochen. 2005. Schlumberger. *Shale Gas White Paper*. 05-OF299. Schlumberger Marketing Communications. October 2005.

[161] Hayden, J., and D. Pursell. 2005. Pickering Energy Partners, Inc. *The Barnett Shale: Visitor' s Guide to the Hottest Gas Play in the US*. October 2005.

[162] Interstate Oil and Gas Compact Commission (IOGCC). *Issues: States' Rights*. http://www.iogcc.state.ok.us/states-rights. Accessed: September 2008.

[163] State Review of Oil and Natural Gas Regulations (STRONGER). http://www.strongerinc.org. Accessed: September 2008.

[164] STRONGER. *List of State Reviews*. http://www.strongerinc.org/reviews/reviews.asp. Accessed: September 2008.

[165] STRONGER. *History of STRONGER - Helping To Make An Experiment Work*. http://www.strongerinc.org/about/history.asp. Accessed: September 2008.

[166] Southlake, Texas, Gas Well Ordinance. Article IV. *Gas and Oil Well Drilling and Production*.

[167] Richard Hills, Texas, Gas Well Ordinance. Ordinance No. 996-04. September 14, 2004.

[168] Haltom City Ordinance No. 0-2004-026-15. November 22, 2004.

[169] Fort Worth, Texas, Ordinance No. 16986-06-2006. June 21, 2006.

[170] Susquehanna River Basin Commission. *Regulation of Projects*. 18 CFR 801, 806, 807, and 808. http://srbc.net/policies/docs/srbc_regulation_of_projects.PDF. Effective: February 20, 2007. Accessed: September 2008.

[171] Delaware River Basin Commission Basin. Administrative Manual Part III. Water Quality Regulations. 18 CFR PART 410. http://www.state.nj.us/drbc/regs/WQRegs_071608.pdf. September 12, 2008.

[172] 33 U.S.C. section 1251 *et. Seq.*

[173] EPA. *Summary of the Clean Water Act*. http://www.epa.gov/lawsregs/laws/cwa.html. Accessed: September 2008.

[174] EPA. *Water Quality and Technology-Based Permitting*. http://cfpub.epa.gov/npdes/generalissues/watertechnology.cfm.Accessed : September 2008.

[175] EPA. *Effluent Limitations Guidelines and Standards*. http://cfpub.epa.gov/npdes/techbasedpermitting/effguide.cfm.Accessed: September 2008.

[176] National Archives and Records Administration. Electronic Code of Federal Regulations. *Protection of the Environment. Oil and Gas Extraction Point Source Category*. 40 CFR Part 435. http://ecfr.gpoaccess.gov/cgi/t/text/textidx?c=ecfr&rgn=div5&view=text&node=40:29.0.1.1.11&idno=40. Effective: June 12, 2006. Accessed: September 2008.

[177] EPA. *Final Rule: Amendments to the Storm Water Regulations for Discharges Associated with Oil and Gas Construction Activities*. http://www.epa.gov/npdes/regulations/final_oil_gas_factsheet.pdf. Effective: June 12, 2006. Accessed: September2008.

[178] *NRDC v. EPA*. 9th Cir. p.5947. 2008.
[179] EPA . *Final Rule: Amendments to the Storm Water Regulations for Discharges Associated with Oil and Gas Construction Activities*. http://www.epa.gov/npdes/regulations/final_oil_gas_factsheet.pdf. Effective June 12, 2006. Accessed: September 2008.
[180] Independent Petroleum Association of America (IPAA). 2004. Guidance Document: *Reasonable and Prudent Practices for Stabilization (RAPPS) Of Oil and Gas Construction Sites*. Prepared by Horizon Environmental Services, Inc. April 2004.
[181] EPA. 2004. *Understanding the Safe Drinking Water Act*. EPA 816-F-04-030. http://www.epa.gov/safewater/sdwa/30th/factsheets/understand.html. June 2004.
[182] EPA. Drinking Water Academy (DWA). 2003. *Introduction to the Underground Injection Control Program*. http://www.epa.gov/safewater/dwa/electronic/presentations/uic/uic.pdf. January 2003.
[183] EPA. *US EPA's Program to Regulate the Placement of Waste Water and other Fluids Underground*. EPA 816-F-04-040. June 2004.
[184] EPA. Underground Injection Control Program. *UIC Program Primacy*. http://www.epa.gov/OGWDW/uic/primacy.html#who. Accessed: September 2008.
[185] EPA. Region 9: Superfund. *Oil Program*. http://www.epa.gov/region09/waste/sfund/oilpp/. Accessed: September 2008.
[186] Ibid.
[187] EPA. Spill Prevention, Control, and Countermeasure (SPCC). *Proposed Rule Amendments, Information about October 2007 proposed amendments to the SPCC Rule*. http://www.epa.gov/emergencies/content/spcc/index.htm. Accessed: September 2008.
[188] EPA. *Water Quality Standards. State, Tribal & Territorial Standards*. http://www.epa.gov/waterscience/standards/wqslibrary/. Accessed: September 2008.
[189] SRBC. http://www.srbc.net/. Accessed: September 2008.
[190] DRBC. http://www.state.nj.us/drbc/. Accessed: September 2008.
[191] Clean Air Act. http://www.epa.gov/air/caa/. Amended in: 1990. Accessed: September 2008.
[192] EPA. *The Green Book Nonattainment Areas for Criteria Pollutants*. http://www.epa.gov/oar/oaqps/greenbk/. Accessed: September 2008.
[193] EPA. Air Trends. *Basic Information*. http://www.epa.gov/airtrends/sixpoll.html. Accessed: September 2008.
[194] Texas Commission on Environmental Quality (TCEQ). *Standard of Performance for New Stationary Sources, Spark Ignition Internal Combustion Engines*. 40 CFR Part 60, Subpart JJJJ.
[195] Texas Commission on Environmental Quality (TCEQ). *National Emission Standard for Hazardous Air Pollutants for Stationary Reciprocating Internal Combustion Engines*. 40 CFR Part 63, Subpart ZZZZ.
[196] EPA. 2007. *Final Emission Standards of Performance for Stationary Spark Ignition Internal Combustion Engines; and Final Air Toxics Standards for Reciprocating Internal Combustion Engines*. http://www.epa.gov/ttn/oarpg/t3pfpr.html. December 20,2007.
[197] EPA. Region2. *Solid Waste; RCRA Subtitle D*. http://www.epa.gov/region2/waste/dsummary.htm. Accessed: September 2008.
[198] EPA. 2002. *Exemption of Oil and Gas Exploration and Production Wastes from Federal Hazardous Waste Regulations*. http://epa.gov/epawaste/nonhaz/industrial/special/oil/oil-gas.pdf. October 2002.
[199] Public Law 96-482, Sec. 1. Oct. 21, 1980. 94 Stat. 2334.
[200] Federal Register Volume 58, Number 5. EPA. 1993. *Clarification of the Regulatory Determination for Wastes From the Exploration, Development and Production of Crude Oil, Natural Gas and Geothermal Energy*. pp 15284 – 15287. March 22, 1993.
[201] Federal Register Volume 53. EPA. 1988. *Regulatory Determination for Oil and Gas and Geothermal Exploration, Development and Production Wastes*. pp 25447- 25458. July 6, 1988.
[202] EPA. 2002. *Exemption of Oil and Gas Exploration and Production Wastes from Federal Hazardous Waste Regulations*. http://epa.gov/epawaste/nonhaz/industrial/special/oil/oil-gas.pdf. October 2002.
[203] EPA. 53 FR 25447. *Regulatory Determination for Oil and Gas and Geothermal Exploration, Development and Production Wastes*. http://epa.gov/osw/nonhaz/industrial/special/oil/og88wp.pdf. Accessed: September 2008.
[204] EPA. *Laws, Regulations, Guidance and Dockets. Summary of the Oil Pollution Act*. 33 U.S.C. Section 2702 et seq. (1990). http://www.epa.gov/lawsregs/laws/opa.html. Accessed: September 2008.
[205] STRONGER. *Process Success and Breakdown*. http://www.strongerinc.org/about/success.asp. Accessed: September 2008.
[206] National Archives and Records Administration. Electronic Code of Federal Regulations. Wildlife and Fisheries. *Endangered and Threatened Wildlife and Plants*. 50 CFR Part 17. http://ecfr.gpoaccess.gov/cgi/t/text/textidx?c=ecfr&tpl=/ecfrbrowse/Title50/50cfr17_main_02.tpl. Accessed: September 2008.
[207] *Babbitt v. Sweet Home Chapter of Communities for a Great Oregon*. 515 U.S. 687. 1995.
[208] Christman, J., and V. Albrecht. Hunton and Williams, LLP. http://library.findlaw.com/1999/Jan/1/241467.html. 1999.

[209] ALL Consulting. 2004. *Coal Bed Natural Gas Handbook. Resources for the Preparation and Reviews of Project Planning Elements and Environmental Documents.* October 2004.
[210] Construction Industry Compliance Assistance (CICA). *The CICA Center.* http://www.cicacenter.org/index.cfm. Accessed:September 2008.
[211] Ibid.
[212] U.S. Fish & Wildlife Service (USFS). Endangered Species Program. *Endangered Species Related Laws, Regulations, Policies & Notices.* http://www.fws.gov/Endangered/policy/index.html. Accessed: September 2008.
[213] Ibid.
[214] Construction Industry Compliance Assistance (CICA). *The CICA Center.* http://www.cicacenter.org/index.cfm. Accessed: September 2008.
[215] ALL Consulting. 2002. *Handbook on Best Management Practices and Mitigation Strategies for Coal Bed Methane in the Montana Portion of the Powder River Basin.* 2002.
[216] EPA. 1998. *RCRA: Superfund & EPCRA Hotline Training Module.* EPA 540-R-98-022. June 1998.
[217] IOGCC. 1996. *Review of Existing Reporting Requirements for Oil and Gas Exploration and Production Operators in Five Key States.* 1996.
[218] IPAA. 2000. Environment & Safety Fact Sheets. *IPAA Opposes EPA's Possible Expansion of TRI.* http://www.ipaa.org/issues/factsheets/environment_safety/tri.asp. May 2000.
[219] U.S. Department of Labor. Occupational Safety & Health Administration. www.osha.gov. Accessed: September 2008.
[220] U.S. Department of Labor. Occupational Safety & Health Administration. *Oil and Gas Well Drilling, Servicing and Storage Standards.* http://www.osha.gov/SLTC/oilgaswelldrilling/standards.html. Accessed: September 2008.
[221] Arkansas Oil and Gas Commission. *Mission Statement. http://www.aogc.state.ar.us/mission.pdf.* Accessed: September 2008.
[222] Ohio Department of Natural Resources, Oil and Gas Division. *Ohio Revised Code 1509.*
[223] ALL Consulting. 2004. *Coal Bed Natural Gas Handbook. Resources for the Preparation and Reviews of Project Planning Elements and Environmental Documents.* October 2004.
[224] Bureau of Land Management (BLM). 2006. *Scientific Inventory of Onshore Federal Lands' Oil and Gas Resources and the Extent and Nature of Restrictions or Impediments to Their Development.* Prepared by the U.S. Departments of the Interior, Agriculture, and Energy. 2006.
[225] Bonner, T., and L. Willer. 2005. National Association of Royalty Owners, Inc (NARO). The New Generation. *Royalty Management 101: Basics for Beginners and a Refresher Course for Everyone.* 25th Anniversary Convention in Oklahoma City, OK. November 3-5, 2005.
[226] Lantz, G. 2008 .The Barnett Shale Magazine. *Drilling Green Along Trinity Trails.* Summer 2008.
[227] Harper, J. 2008. *The Marcellus Shale – An Old "New" Gas Reservoir in Pennsylvania.* Pennsylvania Geology. v 28, no 1. Spring 2008. Published by the Bureau of Topographic and Geologic Survey, Pennsylvania Department of Conservation and Natural Resources.
[228] Frantz, J.K., and V. Jochen. 2005. Schlumberger. *Shale Gas White Paper.* 05-OF299. Schlumberger Marketing Communications. October 2005.
[229] Halliburton Energy Services. 2008. *U.S. Shale Gas: An Unconventional Resource.* Unconventional Challenges. 2008.
[230] Harper, J. 2008. *The Marcellus Shale – An Old "New" Gas Reservoir in Pennsylvania.* Pennsylvania Geology. v 28, no 1. Spring 2008. Published by the Bureau of Topographic and Geologic Survey, Pennsylvania Department of Conservation and Natural Resources.
[231] Halliburton Energy Services. 2008. *U.S. Shale Gas: An Unconventional Resource.* Unconventional Challenges. 2008.
[232] Harper, J. 2008. *The Marcellus Shale – An Old "New" Gas Reservoir in Pennsylvania.* Pennsylvania Geology. v 28, no 1. Spring 2008. Published by the Bureau of Topographic and Geologic Survey, Pennsylvania Department of Conservation and Natural Resources.
[233] Ibid.
[234] Marshall Miller & Associates, Inc. *Marcellus Shale.* Presented to: Fireside Pumpers in Bradford, PA.
[235] Parshall, J. 2008. *Barnett Shale Showcases Tight-gas Development.* Journal of Petroleum Technology. September 19, 2008.
[236] U.S. Department of the Interior. 2008. *Reasonably Foreseeable Development Scenario for Fluid Minerals: Arkansas.* Prepared for the Bureau of Land Management Eastern States Jackson Field Office. March 2008.
[237] Bromley, M. 1985. U.S. Department of Agriculture, Forest Service, Intermountain Research Station. Ogden, Utah. *Wildlife Management Implications of Petroleum Exploration and Development in Wildland Environments.* General Technical Report INT-199. 1985.
[238] Catskill Mountainkeeper. *The Marcellus Shale – America's next super giant.* http://catskillmountainkeeper.org/node/290. Accessed: October 2008.

[239] Venesky, T. 2008. *State-Owned Parcels Eyed for Gas Deposits*. The Times Leader. http://www.timesleader.com/news/hottopics/shale/20080304_01gas_wells_ART.html. October 14, 2008. Accessed: September 2008.

[240] United States Department of the Interior and United States Department of Agriculture. 2007. *Surface Operating Standards and Guidelines for Oil and Gas Exploration and Development*. BLM/WO/ST-06/021+3071/REV07. Bureau of Land Management. Denver, Colorado. 2007.

[241] ALL Consulting. 2004. Coal Bed Natural Gas Handbook. *Resources for the Preparation and Reviews of Project Planning Elements and Environmental Documents*. October 2004.

[242] The Perryman Group. 2007. *Bounty from Below: The Impact of Developing Natural Gas Resources Associated with the Barnett Shale on Business Activity in Fort Worth and the Surrounding 14-County Area*. May 2007.

[243] Satterfield, J., M. Mantell, D. Kathol, F. Hiebert, K. Patterson, and R. Lee. 2008. Chesapeake Energy Corp. *Managing Water Resource's Challenges in Select Natural Gas Shale Plays*. Presented at the GWPC Annual Meeting. September 2008.

[244] Douglas, D. *Anger over road damage caused by Barnett Shale development*. http://www.wfaa.com/sharedcontent/dws/wfaa/latestnews/stories/wfaa080822_lj_douglas.18a895d.html. Accessed: October 2008.

[245] The Perryman Group. 2007. *Bounty from Below: The Impact of Developing Natural Gas Resources Associated with the Barnett Shale on Business Activity in Fort Worth and the Surrounding 14-County Area*. May 2007.

[246] Parshall, J. 2008. *Barnett Shale Showcases Tight-gas Development*. Journal of Petroleum Technology. September 19, 2008.

[247] Various City Ordinances including Arlington ON2006-015, Burleson B-710, Cleburne Chapter 118, Crowley Chapter 37, Fort Worth ON. 16986-06-2006, and Richland Hills, ON.996-04.

[248] Ibid.

[249] Chesapeake Energy Corporation. 2008. *Drilling 101*. Presented to the NY DEC. September 2008.

[250] ALL Consulting. 2002. *Handbook on Best Management Practices and Mitigation Strategies for Coal Bed Methane in the Montana Portion of the Powder River Basin*. 2002.

[251] ALL Consulting. 2008. *Casing and Cement Programs*. Horizontal Well Diagram.

[252] Bellabarba, M., H. Bulte-Loyer, B. Froelich, S. Roy-Delage, R. Van Kuijk, S. Zeroug, D. Guillot, N. Moroni, S. Pastor, and A. Zanchi. 2008. *Ensuring Zonal Isolation Beyond the Life of the Well*. Oilfield Review. Spring 2008.

[253] Railroad Commission of Texas (RRC). http://www.rrc.state.tx.us/. Accessed: September 2008.

[254] Bellabarba, M., H. Bulte-Loyer, B. Froelich, S. Roy-Delage, R. Van Kuijk, S. Zeroug, D. Guillot, N. Moroni, S. Pastor, and A. Zanchi. 2008. *Ensuring Zonal Isolation Beyond the Life of the Well*. Oilfield Review. Spring 2008.

[255] Arkansas Oil and Gas Commission. *General Rules and Regulations*. Accessed: September 2008.

[256] Michie & Associates. 1988. *Oil and Gas Water Injection Well Corrosion*. Prepared for the American Petroleum Institute. 1988.

[257] Ibid.

[258] Michie, T.W., and C.A. Koch. *Evaluation Of Injection-Well Risk Management In The Williston Basin*, June, 1991.

[259] Michie & Associates. 1988. *Oil and Gas Water Injection Well Corrosion*. Prepared for the American Petroleum Institute. 1988.

[260] Schlumberger. *Drilling Fluids: The Many Roles of Drilling Fluids*. http://www.seed.slb.com/en/scictr/watch/mud/char.htm.Accessed: September 2008.

[261] Singh,, M. M., Jr. The Pennsylvania State University. Goodwin, Robert J. Gulf Research and Development Company. 1965. *Mechanism of Drilling Wells with Air as the Drilling Fluid*. SPE 1052-MS.

[262] Kennedy, J. *Technology Limits Environmental Impact of Drilling*. Drilling Contractor. July/August 2000. 33-35.

[263] Swaco, M. 2006. *Reclaim Technology: The system that extends the life of oil- and synthetic-base drilling fluids while reducing disposal and environmental costs*. Published by M-I LLC. 2006.

[264] Oklahoma Department of Environmental Quality. *Pollution Prevention Case Study for Oxy USA, Inc.* http://www.deq.state.ok.us/CSDnew/P2/Casestudy/oxyusa~1.htm. Accessed: September 2008.

[265] Harper, J. 2008. *The Marcellus Shale – An Old "New" Gas Reservoir in Pennsylvania*. Pennsylvania Geology. v 28, no 1. Spring 2008. Published by the Bureau of Topographic and Geologic Survey, Pennsylvania Department of Conservation and Natural Resources..

[266] Veatch, R.W., Jr., Z. A. Moschovidis, and C. R. Fast. *An Overview of Hydraulic Fracturing*. Recent Advances in Hydraulic Fracturing, Edited by J.L. Gidley, S.A. Holditch, D.E. Nierode, and R.W. Veatch Jr. Society of Petroleum Engineers, Henry L Doherty Series Monograph. v 12.

[267] Jennings, A.R., Jr., and W. G. Darden. 1979. *Gas Well Stimulation in the Eastern United States*. SPE 7914. 1979.

[268] Veatch, R.W., Jr., Z. A. Moschovidis, and C. R. Fast. *An Overview of Hydraulic Fracturing*. Recent Advances in Hydraulic Fracturing, Edited by J.L. Gidley, S.A. Holditch, D.E. Nierode, and R.W. Veatch Jr. Society of Petroleum Engineers, Henry L Doherty Series Monograph. v 12.

[269] Boyer, C., J. Kieschnick, R. Suarez-Rivera, R. Lewis, and G. Walter. 2006. Schlumberger. Oilfield Review. *Producing Gas from Its Source*. Pp 36-49. Autumn 2006.

[270] Meyer & Associates, Inc. *User's Guide for the Meyer Fracturing Simulators.* Sixth Edition.

[271] Schlumberger. *PowerSTIM..* http://www.slb.com/content/services/stimulation/fracturing/powerstim.asp? Accessed:September 2008.

[272] Mapping of Microseismic Events Courtesy of: Oilfield Service Company.

[273] Meyer & Associates, Inc. *User's Guide for the Meyer Fracturing Simulators.* Sixth Edition.

[274] Parshall, J. 2008. *Barnett Shale Showcases Tight-gas Development*. Journal of Petroleum Technology. September 19, 2008.

[275] Overbey, W.K., A.B. Yost, and D.A. Wilkins. 1988. *Inducing Multiple Hydraulic Fractures from a Horizontal Wellbore.* SPE Paper 18249.

[276] Chesapeake Energy Corporation. *Components of Hydraulic Fracturing*, Presented to the NY DEC. October 2008.

[277] Arthur, J.D., B. Bohm, and M. Layne. 2008. ALL Consulting. *Hydraulic Fracturing Considerations for Natural Gas Wells of the Marcellus Shale*. Presented at the GWPC Annual Forum in Cincinnati, OH. September 2008.

[278] Ibid.

[279] Harper, J. 2008. *The Marcellus Shale – An Old "New" Gas Reservoir in Pennsylvania*. Pennsylvania Geology. v 28, no 1. Spring 2008. Published by the Bureau of Topographic and Geologic Survey, Pennsylvania Department of Conservation and Natural Resources..

[280] Cramer, D. *Stimulating Unconventional Reservoirs: Lessons Learned, Successful Practices, Areas for Improvement*. SPE 114172.

[281] Harper, J. 2008. *The Marcellus Shale – An Old "New" Gas Reservoir in Pennsylvania*. Pennsylvania Geology. v 28, no 1. Spring 2008. Published by the Bureau of Topographic and Geologic Survey, Pennsylvania Department of Conservation and Natural Resources..

[282] Schlumberger. *Fracturing*. http://www.slb.com/content/services/stimulation/fracturing/index.asp? Accessed: September 2008.

[283] Ketter, A.A., J. L. Daniels, J. R. Heinze, and G. Waters. *A Field Study Optimizing Completion Strategies for Fracture Initiation in Barnett Shale Horizontal Wells*. SPE 103232.

[284] Schlumberger. *PowerSTIM.* http://www.slb.com/content/services/stimulation/fracturing/powerstim.asp? Accessed: September 2008.

[285] Compiled from Data collected at a Fayetteville Shale Fracture Stimulation by ALL Consulting 2008.

[286] Modified from: Arthur, J.D., B. Bohm, and M. Layne. 2008. ALL Consulting. *Hydraulic Fracturing Considerations for Natural Gas Wells of the Marcellus Shale*. Presented at the GWPC Annual Forum in Cincinnati, OH. September 2008.

[287] Satterfield, J., M. Mantell, D. Kathol, F. Hiebert, K. Patterson, and R. Lee. 2008. Chesapeake Energy Corp. *Managing Water Resource's Challenges in Select Natural Gas Shale Plays*. Presented at the GWPC Annual Meeting. September 2008.

[288] Personal conversations with various operators. September 2008.

[289] NRCS. National Water & Climate Center. http://www.wcc.nrcs.usda.gov/. Accessed: September 2008.

[290] Satterfield, J, M. Mantell, D. Kathol, F. Hiebert, K. Patterson, and R. Lee. 2008. Chesapeake Energy Corp. *Managing Water Resource's Challenges in Select Natural Gas Shale Plays*. Presented at the GWPC Annual Meeting. September 2008.

[291] Gaudlip, A., L. Paugh (SPE, Range Resources Appalachia LL), and T. Hayes (Gas Technology Institute). *Marcellus Shale Water Management Challenges in Pennsylvania*. Presented at the SPE Shale Gas Production Conference, November 2008.

[292] Chesapeake Energy Corporation. *Little Red River Project*. Presentation to Trout Unlimited. May 6, 2008.

[293] Weston, R.T. 2008. *Development of the Marcellus Shale-Water Resource Challenges.* Published by Kirkpatrick & Lockhart Preston Gates Ellis LLP. 2008.

[294] Satterfield, J., M. Mantell, D. Kathol, F. Hiebert, K. Patterson, and R. Lee. . 2008. Chesapeake Energy Corp. *Managing Water Resource's Challenges in Select Natural Gas Shale Plays*. Presented at the GWPC Annual Meeting. September 2008.

[295] Personal communication with numerous operator and service companies in a variety of shale gas plays.

[296] Willberg, D.M., N. Steinsberger, R. Hoover, R. J. Card, and J. Queen. 1998. *Optimization of Fracture Cleanup Using Flowback Analysis*. SPE 39920-MS. 1998.

[297] Satterfield, J., M. Mantell, D. Kathol, F. Hiebert, K. Patterson, and R. Lee. *Managing Water Resource's Challenges in Select Natural Gas Shale Plays*. Chesapeake Energy Corp. Presented at the GWPC Annual Meeting. September 2008.

[298] Personal conversations with various operators. March, 2009.

[299] Satterfield, J., Mantell, M., Kathol, D., Hiebert, F., Patterson, K., and Lee, R. 2008. Chesapeake Energy Corp. *Managing Water Resource's Challenges in Select Natural Gas Shale Plays*. Presented at the GWPC Annual Meeting. September 2008.

[300] Harper, J. 2008. *The Marcellus Shale – An Old "New" Gas Reservoir in Pennsylvania*. Pennsylvania Geology. v 28, no 1. Spring 2008. Published by the Bureau of Topographic and Geologic Survey, Pennsylvania Department of Conservation and Natural Resources.

[301] Railroad Commission of Texas (RRC). v2008.v *Water Use in the Barnett Shale*. vhttp://www.rrc.state.tx.us/divisions/og/wateruse_barnettshale.html. vUpdated: vJune 30, 2008. vAccessed: vOctober 2008.

[302] Personal communication with a variety of producing and service companies operating in shale gas basins. 2008.

[303] Burnett, D.B., and C. J. Vavra. 2006. Global Petroleum Research Institute (GPRI) and Food Protein Research Center. Texas A&M Produced Water Treatment. *Desalination of Oilfield Brine. New Resources: Produced Water Desalination*. August 2006.

[304] Grable, R.C. 2008. Kelly Hart & Hallman, Attorneys at Law. *Saltwater Disposal and Other "Hot Issues" in Urban Drilling. Safety-Wellbore*. Presented at the 19th Annual State Bar of Texas Advanced Real Estate Drafting Course in Dallas. February 29, 2008.

[305] Burnett, D.B., and C. J. Vavra. 2006. Global Petroleum Research Institute (GPRI) and Food Protein Research Center. Texas A&M Produced Water Treatment. *Desalination of Oilfield Brine. New Resources: Produced Water Desalination*. August 2006.

[306] Ibid.

[307] University of Arkansas. Argonne National Laboratory. *Fayetteville Shale Natural Gas: Reducing Environmental Impacts. Well Production and Water Management*. http://lingo.cast.uark.edu/LINGOPUBLIC/index.htm. Accessed: October 2008.

[308] Ibid.

[309] University of Arkansas. Argonne National Laboratory. *Fayetteville Shale Natural Gas: Reducing Environmental Impacts. Well Production and Water Management*. http://lingo.cast.uark.edu/LINGOPUBLIC/environ/wellprod/index.htm. Accessed:

[310] Darbonne, N. Oil and Gas Investor. *Engelder: Marcellus, Mona Lisa Share Structural-Fracturing Features*. http://www.oilandgasinvestor.com/Headlines/WebJuly/item5298.php. July 18, 2008.

[311] Personal communication with Pennsylvania Department of Environmental Protection, Bureau of Oil and Gas Management. October 2008.

[312] Personal communication with operators and service industry personnel in the Pennsylvania area. October 2008.

[313] University of Arkansas. Argonne National Laboratory. *Fayetteville Shale Natural Gas: Reducing Environmental Impacts. Well Production and Water Management*. http://lingo.cast.uark.edu/LINGOPUBLIC/environ/wellprod/index.htm. Accessed: October 2008.

[314] Personal communication with Tim Baker from Oklahoma Corporation Commission, Oil and Gas Conservation Division. October 16, 2008.

[315] Ibid.

[316] Ibid.

[317] Burnett, D.B., and C. J. Vavra. 2006. Global Petroleum Research Institute (GPRI) and Food Protein Research Center. Texas A&M Produced Water Treatment. *Desalination of Oilfield Brine. New Resources: Produced Water Desalination*. August 2006.

[318] Grable, R.C. 2008. Kelly Hart & Hallman, Attorneys at Law. Saltwater Disposal and Other "Hot Issues" in Urban Drilling. Safety-Wellbore. Presented at the 19th Annual State Bar of Texas Advanced Real Estate Drafting Course in Dallas. February 29, 2008.

[319] Ewing, J. 2008. Devon Energy Corp. *Taking a Proactive Approach to Water Recycling in the Barnett Shale*. Presented at the Fort Worth Business Press Barnett Shale Symposium. February 29, 2008.

[320] Texas Railroad Commission. *NORM – Naturally Occurring Radioactive Material*. http://www.rrc.state.tx.us/divisions/og/key-programs/norm.html. Accessed: October 2008.

[321] NCRP. 2009. *Ioning Radiation Exposure of the Populations of the United States*. National Council on Radiation Exposure and Measurements, report No. 160. March, 2009.

[322] USGS. 1999. *Naturally Occurring Radioactive Materials (NORM) in Produced Water and Oil-Field Equipment – An Issue for the Energy Industry*. USGS Fact Sheet FS-142-99. September 1999.

[323] Smith, K.P., D. L. Blunt, G. P. Williams, and C. L. Tebes. 1996. Argonne National Laboratory. Environmental Assessment Division. *Radiological Dose Assessment Related to Management of Naturally Occurring Radioactive Materials Generated by the Petroleum Industry*. September 1996.

[324] Ibid.

[325] API. 2004. *Naturally Occurring Radioactive Material in North American Oilfields* A Fact Sheet from the American Petroleum Institute. 2004.

[326] Barnett Shale Energy Education Council. *Environment*. www.bseec.org/index.php/content/facts/environment. Accessed: September 2008.

[327] Ibid.

[328] Smith, K.P., D. L. Blunt, G. P. Williams, and C. L. Tebes. 1996. Argonne National Laboratory. Environmental Assessment Division. *Radiological Dose Assessment Related to Management of Naturally Occurring Radioactive Materials Generated by the Petroleum Industry*. September 1996.

[329] Ibid.

[330] NaturalGas.org. *Natural Gas and the Environment*. www.naturalgas.org/environment/naturalgas.asp. Accessed: September 2008.

[331] EPA. *Emissions by Category Chart – Criteria Air Pollutants*. Volatile Organic Compounds - 2001. www.epa.gov/oar/data/emcatbar.html?us~USA~United%20States. Accessed: September 2008.

[332] EPA. Technology Transfer Network: 1999 National-Scale Air Toxics Assessment. *Benzene Emissions 1999*. http://www.epa.gov/ttn/atw/nata1999/99pdfs/benzene.pdf. Accessed: September 2008.

[333] EPA. *Emissions by Category Chart – Criteria Air Pollutants*. Carbon Monoxide – 2001. www.epa.gov/oar/data/emcatbar.html?us~USA~United%20States. Accessed: September 2008.

[334] Ibid.

[335] Triplepundit. 2008. Shale Gas: *Energy Boon or Environmental Bane, Part One* . www.triplepundit.com/pages/shale-gasenergy-boon-or-envir-003396.php. August 11, 2008.

[336] American Petroleum Institute (API). 2001. RP 49,. *Recommended Practice for Drilling and Well Servicing Operations Involving Hydrogen Sulfide*. 2nd Edition. May 2001. Reaffirmed, March 2007.

[337] EPA, 2008. *Partner Update: Partner Profile: ConocoPhillips Makes New Commitment to Natural Gas Star*. www.epa.gov/gasstar/pdf/ngspartnerup_spring2007.pdf. Spring 2008.

[338] Ibid.

[339] NaturalGas.org. *Processing Natural Gas*. www.naturalgas.org/naturalgas/processing_ng.asp. Accessed: September 2008.

[340] EPA. 2005. *Reduced Emission Completions (Green Completions)*. www.epa.gov/gasstar/workshops/houstonoct26/green_c.pdf. October 26, 2005.

[341] Methane to Markets. *About the Partnership*. www.methanetomarkets.org/about/index.htm. Accessed: September 2008.

[342] Methane to Markets. *Oil and Gas Sector Members*. www.methanetomarkets.org/partners/network/pnmList.aspx?type=oil. Accessed: September 2008.

[343] Methane to Markets. *Methane Partnership Program's*. www.epa.gov/methanetomarkets/partnership.htm. Accessed: September 2008.

In: Shale Gas Development
Editor: Katelyn M. Nash pp. 105-149
ISBN: 978-1-61668-545-4

Chapter 2

UNCONVENTIONAL GAS SHALES: DEVELOPMENT, TECHNOLOGY, AND POLICY ISSUES*

Anthony Andrews[1], Peter Folger[2], Marc Humphries[3], Claudia Copeland[4], Mary Tiemann[5], Robert Meltz[6] and Cynthia Brougher[6]

[1] Energy and Energy Infrastructure Policy
[2] Energy and Natural Resources Policy
[3] Analyst in Energy Policy
[4] Resources and Environmental Policy
[5] Environmental Policy
[6] Legislative Attorney

SUMMARY

In the past, the oil and gas industry considered gas locked in tight, impermeable shale uneconomical to produce. However, advances in directional well drilling and reservoir stimulation have dramatically increased gas production from unconventional shales. The United States Geological Survey estimates that 200 trillion cubic feet of natural gas may be technically recoverable from these shales. Recent high natural gas prices have also stimulated interest in developing gas shales. Although natural gas prices fell dramatically in 2009, there is an expectation that the demand for natural gas will increase. Developing these shales comes with some controversy, though.

The hydraulic fracturing treatments used to stimulate gas production from shale have stirred environmental concerns over excessive water consumption, drinking water well contamination, and surface water contamination from both drilling activities and fracturing fluid disposal.

The saline "flowback" water pumped back to the surface after the fracturing process poses a significant environmental management challenge in the Marcellus region. The

* This is an edited, reformatted and augmented version of a CRS Report for Congress publication dated October 2009.

flowback's high content of total dissolved solids (TDS) and other contaminants must be disposed of or adequately treated before discharged to surface waters. The federal Clean Water Act and state laws regulate the discharge of this flowback water and other drilling wastewater to surface waters, while the Safe Drinking Water Act (SDWA) regulates deep well injection of such wastewater. Hydraulically fractured wells are also subject to various state regulations. Historically, the EPA has not regulated hydraulic fracturing, and the 2005 Energy Policy Act exempted hydraulic fracturing from SDWA regulation. Recently introduced bills would make hydraulic fracturing subject to regulation under SDWA, while another bill would affirm the current regulatory exemption.

Gas shale development takes place on both private and state-owned lands. Royalty rates paid to state and private landowners for shale gas leases range from 12½ % to 20%. The four states (New York, Pennsylvania, Texas, and West Virginia) discussed in this chapter have shown significant increases in the amounts paid as signing bonuses and increases in royalty rates. Although federal lands also overlie gas shale resources, the leasing restrictions and the low resource-potential may diminish development prospects on some federal lands. The practice of severing mineral rights from surface ownership is not unique to the gas shale development. Mineral owners retain the right to access surface property to develop their holdings. Some landowners, however, may not have realized the intrusion that could result from mineral development on their property.

Although a gas-transmission pipeline-network is in place to supply the northeast United States, gas producers would need to construct an extensive network of gathering pipelines and supporting infrastructure to move the gas from the well fields to the transmission pipelines, as is the case for developing any new well field.

BACKGROUND

Until recently, many oil and gas companies considered natural gas locked in tight, impermeable shale uneconomical to produce. Advanced drilling and reservoir stimulation methods have dramatically increased the gas production from "unconventional shales." The Barnett Shale formation in Texas has experienced the most rapid development. The Marcellus Shale formation of the Appalachian basin, in the northeastern United States, potentially represents the largest unconventional gas resource in the United States. Other shale formations, such as the Haynesville shale, straddling Texas and Louisiana, have also attracted interest, as have some formations in Canada. The resource potential of the shales has significantly increased the natural gas reserve estimates in the United States.[1]

Recent high natural gas prices had stimulated interest in gas shales, although prices fell below $4.00 per million Btu (mmBtu) in the summer of 2009.[2] Despite the currently low natural gas prices, there is an expectation that the demand for natural gas will increase and that unconventional gas shales will play a major role in meeting the increased demand for natural gas.

The shale's development is a subject of controversy, however. The potential economic benefits from both the drilling activities and the lease and royalty payments compete with the public's concern for environmentally safe drilling practices and protection of groundwater and surface water.

Directional drilling and "hydraulic fracturing" are instrumental in exploiting this resource. Although oil and gas developers have applied these technologies in conventional oil fields for some time, they have only recently begun applying them to unconventional gas shales. As with oil and gas production generally, gas shale development is primarily subject to state law and regulation, although provisions of two relevant federal laws—the Safe Drinking Water Act (SDWA) and the Clean Water Act (CWA)—also apply. Regulation of well construction differs by state, and federal law currently exempts from regulation the underground injection of fluids for hydraulic fracturing purposes. Two recently introduced bills would subject hydraulic fracturing to regulation under the Safe Drinking Water Act, while a third bill would affirm the current SDWA exemption. There has been an increase in reports of public concerns that hydraulic fracturing poses a potential risk to water wells and thus domestic drinking water supply. The concern is that the chemicals used pose a potential risk to groundwater quality, and the fracturing may damage aquifers. Critics maintain that the large quantities of water that hydraulic fracturing consumes may tax local and regional water supplies and that disposing the "flowback" extracted from the shale after fracturing may affect surface water and groundwater quality. Understanding the technical principles of well drilling, well construction, and stimulation methods are important to assessing the concerns that have arisen over the shale's development.

Gas shale development takes place on both private and state lands. Royalty rates, rents, and signing bonuses paid to state and private landowners vary. In Texas, where similar gas shale development takes place, the rates are generally higher. Although federal lands overlie some portions of the Marcellus, restrictions to drilling and low resource potential may make them less desirable to develop. In the case of split-estate lands, which sever mineral rights from surface ownership, the mineral-owner retains the right to access the land for development. Landowners may not be aware of the mineral-owners' access rights, and are not due compensation for access. Furthermore, they may have to contend with long-term easements for the gathering pipelines that connect the wells to natural gas transmission lines. The states do regulate reclamation of the drilling sites to varying degrees.

This chapter does not discuss all unconventional gas shales. Rather, this chapter limits discussion to the Barnett and Marcellus Shale formations, which are most representative of the resource. Both formations also serve to illustrate the technical and policy issues that are most likely common to developing all gas shales.

Unconventional Gas Shale Resources in the United States

Unconventional gas shales are fine grained, organic rich, sedimentary rocks. The shales are both the source of and the reservoir for natural gas, unlike conventional petroleum reservoirs. In the shales, gas occupies pore spaces, and organic matter adsorbs gas on its surface. The Society of Petroleum Engineers describes "unconventional resources" as petroleum accumulations that are pervasive throughout a large area and that are not significantly affected by hydrodynamic influences (they are also called "continuous-type deposits"). In contrast, conventional petroleum and natural gas occur in porous sandstone and carbonate reservoirs. Under hydrodynamic pressure exerted by water, the petroleum migrated upward from its organic source until an impermeable cap-rock (such as shale) trapped it in the reservoir rock. The "gas-cap" that accumulated over the petroleum has been the source of most produced natural gas.

Though the shales may be as porous as other sedimentary reservoir rocks, their extremely small pore sizes make them relatively impermeable to gas flow, unless natural or artificial fractures occur. Major gas shale basins exist throughout the lower-48 United States. There are at least 21 shale basins in more than 20 states (see **Figure 1**).[3]

Based on a recent assessment of natural gas resources, the United States has a base of 1,836 trillion cubic-feet (tcf).[4] Shale gas made up an estimated one-third of the resource base, roughly 616 tcf. Stated in other terms, shale gas represents the equivalent of approximately 102 billion barrels of crude oil.

Natural gas production in the United States had been declining until recently. After peaking at nearly 24.2 tcf in 2000, production declined to 23.5 tcf by 2005.[5] The decline had reversed by 2008 when production rose to over 26 tcf. (Gulf of Mexico offshore gas production continues to decline, however, and the 2.3 tcf produced in 2008 was only 47% of the 2000 Gulf production level.)

Since 1990, U.S. gas production increased in large part due to steadily increasing production of unconventional sources. The U.S. Geological Survey (USGS) estimates that about 200 tcf of natural gas may be technically recoverable from the shale with a recovery factor of 10% to 1 5%.[6] These sources include shale gas, coal-bed methane, and tight sands. Unconventional gas production increased 65% from 1998 (5.4 tcf/year) to 2007 (8.9 tcf/year).[7]

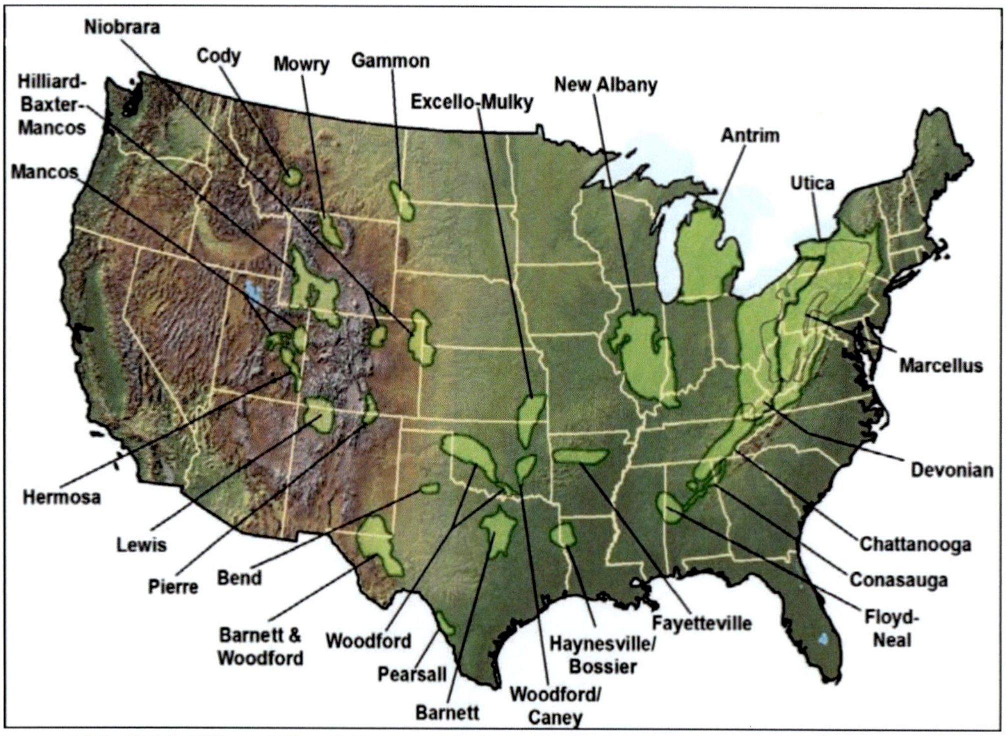

Source: U.S. DOE Office of Fossil Energy, *Modern Shale Gas Development in the United States: A Primer.*

Figure 1. Major Shale Basins in the Conterminous United States

Natural gas production in the "big 7" shale plays (Antrim, Barnett, Devonian, Fayetteville, Woodford, Haynesville, Marcellus) could reach an estimated 27 to 39 bcf/day within 10 to 15 years.[8] Development, however, has been uneven in the Marcellus. Although production data are incomplete at best, reports indicate that Pennsylvania leads the region in producing wells, with West Virginia and Ohio trailing. Navigant expects that the Marcellus basin will be the next significant gas shale play. It appears potentially larger than the other basins already developed.

In 2006, the Bank of America estimated break-even costs for shale gas production to be within a range of \$4.20 to \$11.50 per thousand cubic feet (mcf).[9] The estimated median break-even cost was \$6.64 per mcf. Leading shale gas producers include Chesapeake Energy Corporation, Devon Energy, XTO Energy, Inc., Southwestern Energy, Newfield Exploration Company, and Encana. In late 2008, StatoilHydro, the second largest natural gas supplier to Europe, and Chesapeake Energy Corporation, the largest U.S. natural gas producer, announced a joint agreement to explore unconventional gas opportunities. Under these agreements StatoilHydro will initially acquire a 32.5% interest in Chesapeake's Marcellus Shale gas acreage covering 1.8 million net acres (StatoilHydro's share equals approximately 0.6 million net acres of the leasehold).[10]

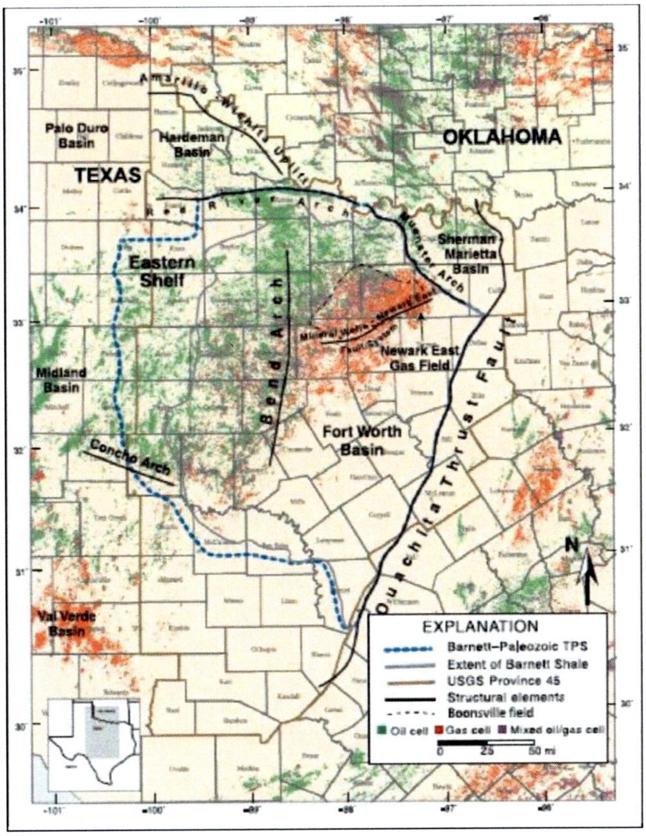

Source: Pollastro et al, *Geologic Framework of the Mississippian Barnett Shale, Barnett-Paleozoic Total Petroleum System, Bend Arch-Fort Worth Basin*, Texas, AAPG Bulletin, V. 91, No. 4 (April 2007), pp 405-436.

Figure 2. Bend Arch-Fort Worth Basin Area

Barnett Shale Formation

The Barnett Shale formation is Mississippian-age black shale that has a high organic content.[11] It underlies 5,000 square miles of the Dallas/Fort Worth area of Texas (primarily in the Fort Worth Basin) to depths of 6,500 to 8,500 feet. (See **Figure 2**.)

Natural Gas Resource Potential

The Barnett Shale play is reportedly the most active natural gas play in the United States with as many as 173 drilling rigs at work in the past year. The USGS estimated that as much as 26.7 tcf of natural gas could be present in continuous accumulations as nonassociated gas trapped in strata of two of the three Barnett Shale Assessment Units (AU)—the Greater Newark East Frac-Barrier Continuous Barnett Shale Gas AU and the Extended Continuous Barnett Shale Gas AU.[12] Collectively, the units as comprise the Bend Arch–Fort Worth Basin.

The area has produced oil and gas since the early 1900s, but mostly from conventional reservoirs.[13] There are over 8,000 wells producing gas from the Barnett formation.[14] Gas production increased from 94 million cubic feet per day (mmcf/day) in 1998 to over 3 billion cubic feet/day (bcf/day) in 2007; an increase of over 3,000%.

The Newark East field of the Barnett has been the largest gas-producing field in Texas since 2000, and ranked as the second highest U.S. gas-producing field in 2005. Gas production rose from less than 11 bcf of natural gas in 1993 to about 480 bcf by 2005.[15] From January 1993 to January 2006, cumulative gas production measured about 1.8 tcf. The Newark East field proven gas reserve estimates range from between 2.5 and 3.0 tcf.

Southwest Regional Gas Supply and Demand

Several major pipeline corridors transport natural gas in the United States. Five major pipeline routes extend from the producing areas in the Southwest, and more than 20 of the major interstate pipelines originate in the Southwest Region.[16] (See **Figure 3**.) In particular, major pipeline networks, constructed after World War II, transport gas from the Gulf of Mexico region to the Northeast, from the Southwest to the Northeast, and from the Southwest to the Midwest.[17] Current pipelines have the capacity to transport as much as 45.2 bcf per day from the region: 62% to the Southeast Region, 20% to the Central Region, 13% to the Western Region, and the rest to Mexico. In response to increased natural gas production and supply, particularly from the Barnett Shale, the Southwest region recently expanded its pipeline infrastructure on a large-scale. In 2008, 30 new pipelines comprising 1,382 miles reached completion; nearly double the previous year's. Thirteen of the new pipelines related to or expanded the northeast Texas area to new development of gas supplies from the Barnett, Woodford, or Fayetteville Shale formations. The remaining pipelines support increased liquefied natural gas (LNG) imports through Texas marine terminals.

Groundwater Resource Issues

As gas development in the Barnett Shale increased over the past decade, water use also increased. One study estimated that water used for Barnett Shale gas development increased from approximately 700 acre-feet (AF) in 2000 to more than 7,000 AF in 2005.[18] Barnett Shale development uses both surface water and groundwater resources, primarily for hydraulically fracturing vertical and horizontal wells. Depending on the well type, from 1.2 to 3.5 million gallons (4 to 11 AF) of water may be consumed in hydraulically fracturing a gas well. Of the approximately 7,000 AF used in 2005, about 60% came from groundwater in the Trinity and Woodbine aquifers in north central Texas.[19] The deeper Trinity aquifer (underlying 61 counties) is much more extensive than the shallower Woodbine aquifer (underlying 17 counties).[20] The state of Texas also considers Trinity a major aquifer compared to the minor Woodbine.[21]

The Trinity Aquifer extends from south-central Texas to southeastern Oklahoma in a 550 mile-long arc-like band.[22] (See **Figure 4.**) It is an important water source for many communities, particularly rural, in north-central Texas where Barnett Shale development is most intense. Groundwater use varies across the Barnett Shale development area; for example, groundwater provides approximately 85% of the total water supply in Cooke County but only 1% for Dallas County.[23] Extensive development of the Trinity Aquifer in the Dallas-Ft. Worth metropolitan areas had caused groundwater levels to drop over 500 feet in some areas. However, when their populations began increasing after the 1970s, local communities abandoned many public supply wells in favor of surface water supplies. This resulted in a recovery of groundwater levels to some extent.[24] For many rural areas, however, groundwater from the Trinity Aquifer remains the sole water source. The Woodbine Aquifer, which is shallower than the Trinity, offers another groundwater supply for the region. The lowest of its three water-bearing layers yields the most water. Though the Woodbine is a minor aquifer, it could be locally important to Barnett Shale development given its proximity.

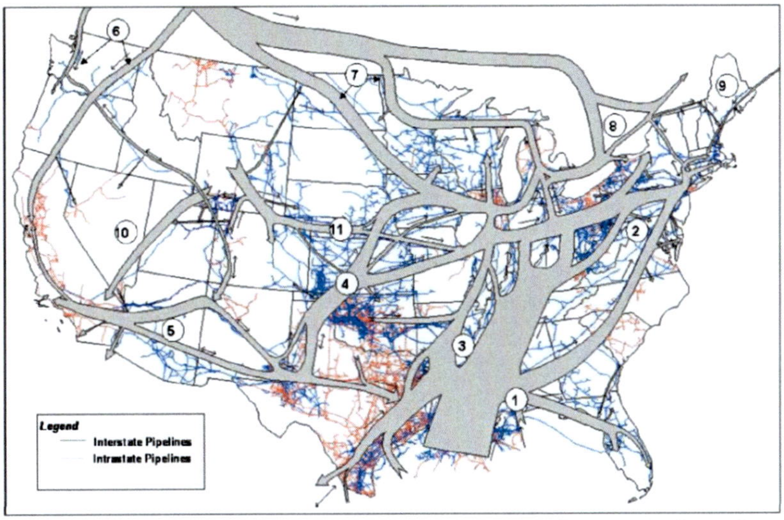

Source: EIA, Gas Tran Gas Transportation Information System http://www.eia.doe.gov/pub/oil_gas/natural_gas/analysis_publications/ngpipeline/transcorr_map.html.

Notes: major pipeline networks originating from the Southwest are: (1) Southwest-Southeast, (2) Southwest- Northeast, (3) Southwest-Midwest, (4) Southwest Panhandle-Midwest, and (5) Southwest-Western. The EIA has determined that this informational map does not raise security concerns, based on the application of the Federal Geographic Data Committee's *Guidelines for Providing Appropriate Access to Geospatial Data in Response to Security Concerns*.

Figure 3. Major Natural Gas Transportation Corridors in the Conterminous United States, 2008

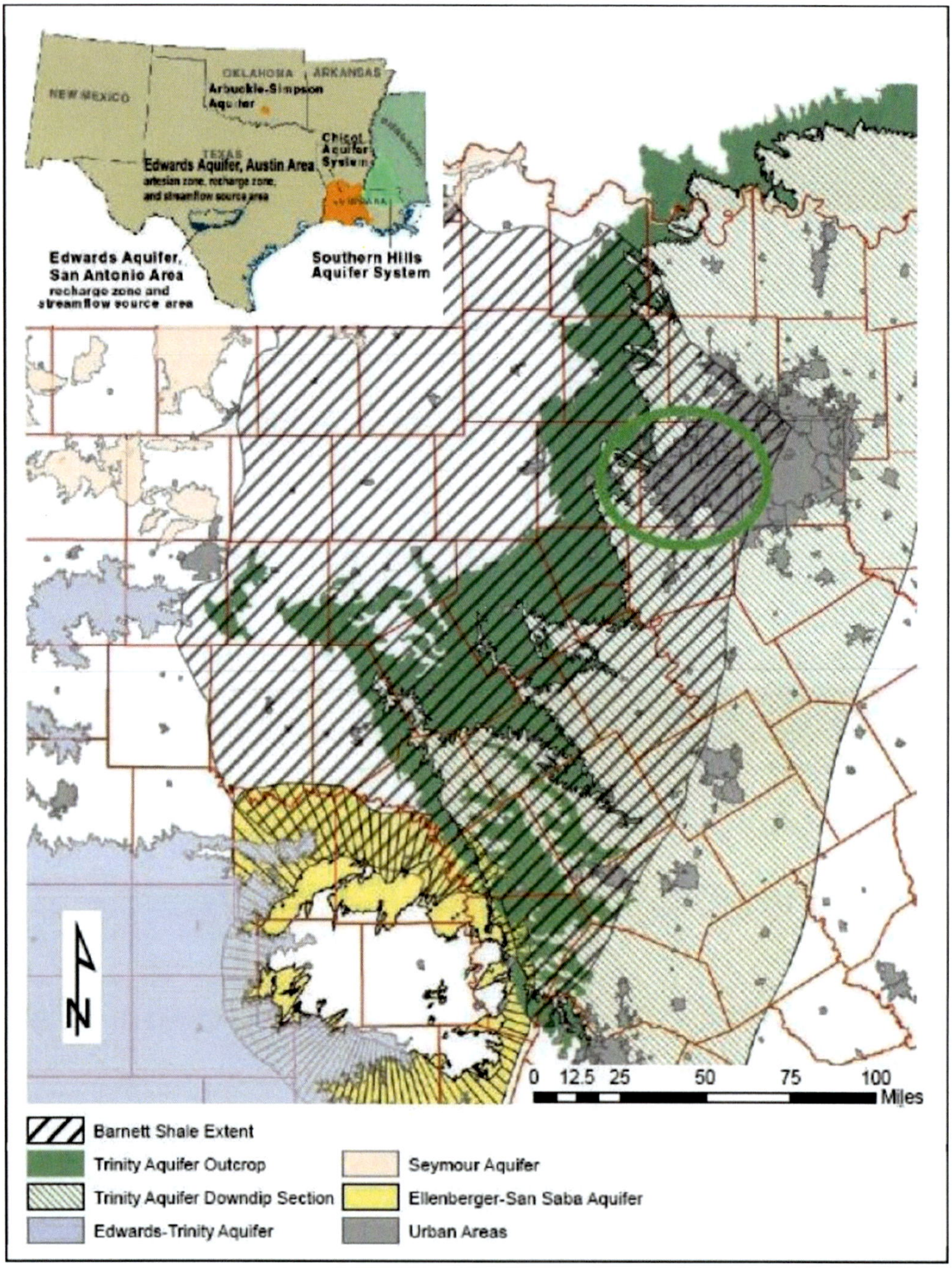

Source: James Bene et al., Northern Trinity/Woodbine Groundwater Availability Model: Assessment of Groundwater Use in the Northern Trinity Aquifer Due to Urban Growth and Barnett Shale Development, R.W. Harden & Associates. p. 2-5 Corner Inset Map: U.S. EPA, *Sole Source Aquifer Protection Program*, Regions VI (modified by CRS) .

Notes: Green circled area represents the so-called "core area" where most of the Barnett Shale gas development has taken place.

Figure 4. Aquifers in the Bend Arch-Fort Worth Basin Area

A study that modeled groundwater use in the area concluded that about 3% of the use was associated with the Barnett Shale's development in 2005.[25] Depending on the pace of shale development, water use (which is essential to hydraulic fracturing) could vary widely.[26] According to the study, low natural gas prices would slow interest in development and thus slightly decrease groundwater use by 2025. High natural gas prices could stimulate accelerated development and thus increase groundwater demand for development to about 10,000 to 25,000 AF by 2025, or from 3% in 2005 to 7% to 13% by 2025.

Increasing shale development could compete with other users for the same groundwater resources, particularly in rural areas where groundwater is a significant fraction of water supplies. It is uncertain how urban areas may react to groundwater well drilling in support of shale development (that is, for hydraulic fracturing), even though groundwater comprises a smaller fraction of the water supply compared to some more rural areas. The modeling study pointed out, as well, that the Trinity and Woodbine aquifers underlie only the eastern portion of the Barnett Shale; no major or minor aquifers underlie the western portions.[27] Development of western portions raises the possibility of transporting groundwater pumped from the Trinity and Woodbine aquifers in the east to the western portion of the Barnett Shale, which could raise estimates of potential groundwater use above those presented in the model results.

Marcellus Shale Formation

The Marcellus Shale is sedimentary rock formation deposited over 350 million years ago during the middle-Devonian period on the geologic timescale. Geologic strata deposited in the Appalachian basin during this period are likely to produce more gas than oil. Regional oil production is associated with Pennsylvanian age strata (of the later Carboniferous period). Most of this black, organic-rich shale lies beneath much of West Virginia, western and northeastern Pennsylvania, southern New York, eastern Ohio, and parts of Virginia and Maryland (see **Figure 5**). It is an estimated 95,000 square miles in areal extent and ranges from 4,000 feet to 8,500 feet in depth.[28] The shale's non-uniform thickness varies from 50 feet to 250 feet as shown by the isopach map of **Figure 5**.[29] Some reports indicate that shale's thickness may be as much as 900 feet in places, however. As shown in **Figure 6**, the Marcellus plunges in depth the further north it goes along the cross section.

Natural Gas Resource Potential

The USGS indicates that the Marcellus Shale may have a mean undiscovered natural gas resource potential of nearly 2 tcf. Considering the total extent of Devonian/Ohio basin shales (which include the Marcellus formation), an estimated 12 tcf may be present, although not all of it may be economically recoverable. (See **Figure 7**.) The USGS estimate, however, is considerably lower than a 2008 "conservative" estimate of 516 tcf made by geoscience professors Terry Engelder (Pennsylvania State University) and Gary Lash (State University of New York). The two professors believe that there could be at least 50 tcf of gas technically recoverable from the Marcellus Shale.[30]

Typically, thicker shales with greater organic material yield more gas, and thus, are more economically desirable to produce. Shale in northeast Pennsylvania and southeast New York has these characteristics and produces dry natural gas. Shale in western Pennsylvania and New York produces a wetter gas that contains petroleum liquids.

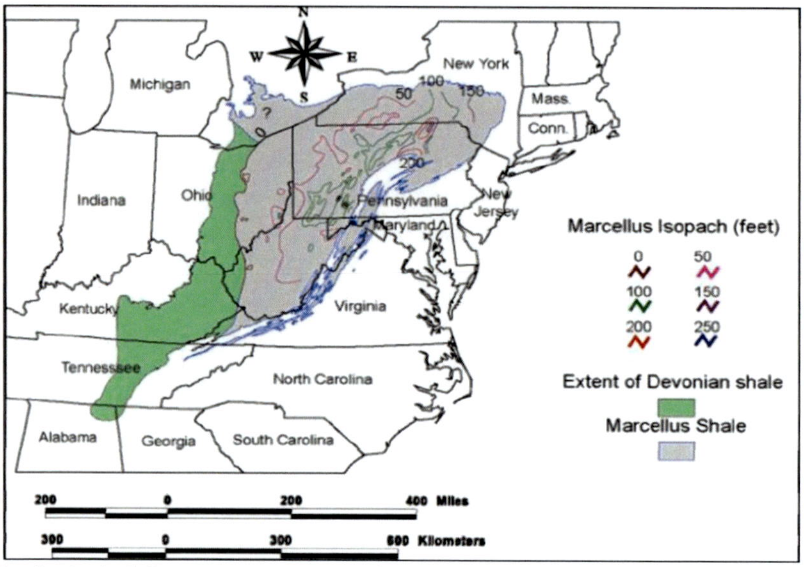

Source: U.S. Geological Survey, *Assessment of Undiscovered Natural Gas Resources in Devonian Black Shales, Appalachian Basin*, Eastern U.S.A., Open-File Report 200-1268, 2005.

Figure 5. Marcellus Shale Formation Thickness

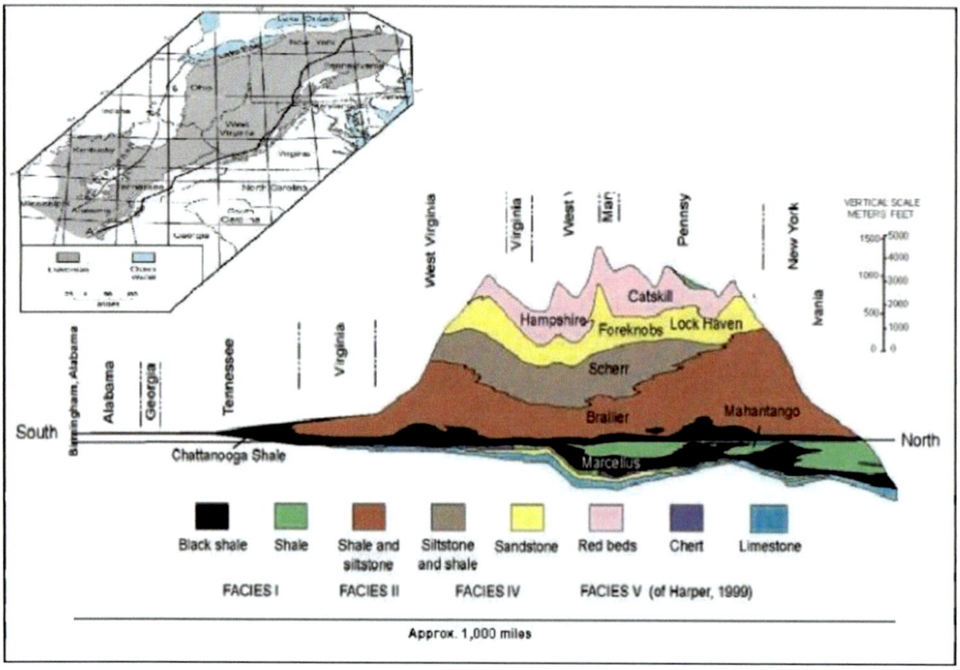

Source: Compiled from USGS Open File Report 200-1268, 2005.
Notes: Cross section from Alabama to New York. Vertical scale is exaggerated.

Figure 6. Devonian Shale Cross Section

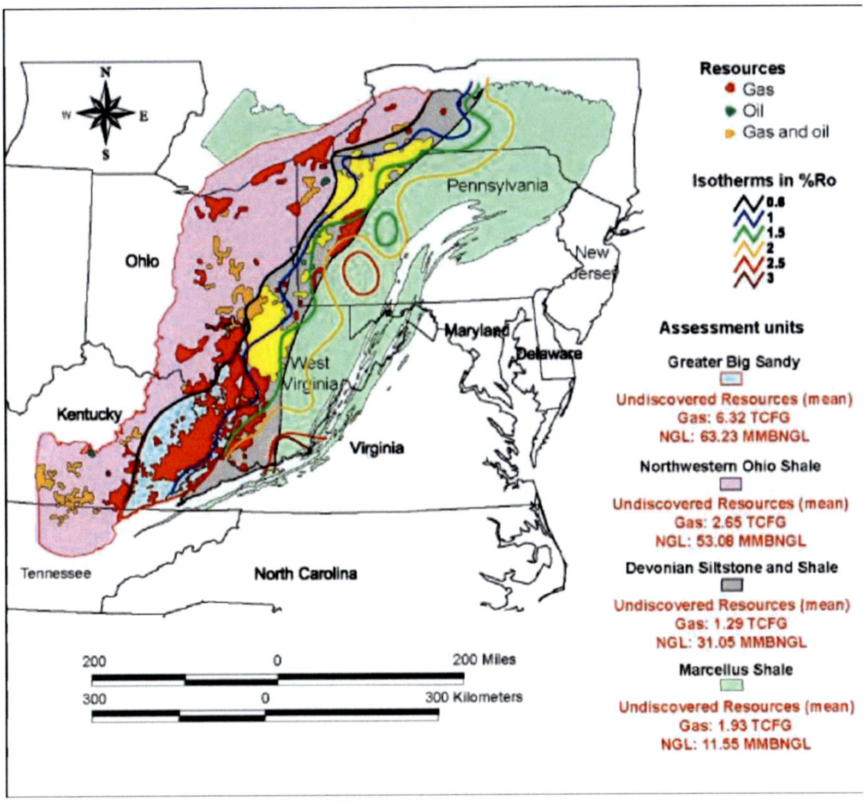

Source: USGS Open File Report 2005-1268.
Notes: NGL: natural gas liquids include butane and propane.

Figure 7. Devonian Shale Undiscovered Resource Potential

StatoilHydro assumes that each well drilled in the Marcellus may have an average estimated ultimate recovery (EUR) of 3.1 bcf.[31] The EUR per well will depend on the length of the horizontal well drilled, the number of fractures in the well, and the quality of the shale. StatoilHydro assumes that each horizontal well would measure 3,000 feet in length with 6 hydraulic fractures intervals in each well. An average well might produce for upwards of 60 years.

Northeast Regional Natural Gas Supply and Demand

In 2007, the northeast region consumed roughly 4 tcf of natural gas.[32] New York led the region in consumption with over 1.19 tcf. The United States as a whole consumed more than 23 tcf.[33] The region produced roughly 580 bcf of natural gas from some 113,000 operating gas wells. Pennsylvania and West Virginia combined made up nearly 89% of the production, with New York and Virginia making up the balance. In summary, the region consumes about seven times as much natural gas as it currently produces. The 50-tcf of gas, estimated by Engelder and Lash as technically recoverable from the Marcellus Shale, would be sufficient to supply the region for 13 years at the current rate of consumption. Taking the entire 51 3-tcf resource-potential into account and assuming that recovery methods would improve might extend the supply further.

Currently, 20 interstate natural gas transmission pipelines serve the northeast region of the United States (see **Figure 8**).[34] This pipeline system delivers natural gas to several intrastate natural gas pipelines and at least 50 local distribution companies in the region. In addition to the natural gas produced in the region, several long-distance natural gas transmission pipelines supply the region from the southeast into Virginia and West Virginia, and from the Midwest into West Virginia and Pennsylvania. Canadian imports come into the region principally through New York, Maine, and New Hampshire. Liquefied natural gas (LNG) supplies also enter the region through import terminals located in Massachusetts, Maryland, and New Brunswick, Canada.

The natural gas produced from the eastern portion of the Marcellus Shale is of high enough quality that it requires little or no treatment for injection into transmission pipelines.[35] A gas transmission pipeline already serves the northeast United States. The Millennium Pipeline project in southern New York could accommodate any increased shale gas production from New York and parts of Pennsylvania to serve the natural gas needs of the region. However, the terrain in West Virginia presents an obstacle to developing additional pipeline capacity and other support infrastructure. Gas producers would also have to construct an extensive network of gathering pipelines to bring the gas out of the well fields.

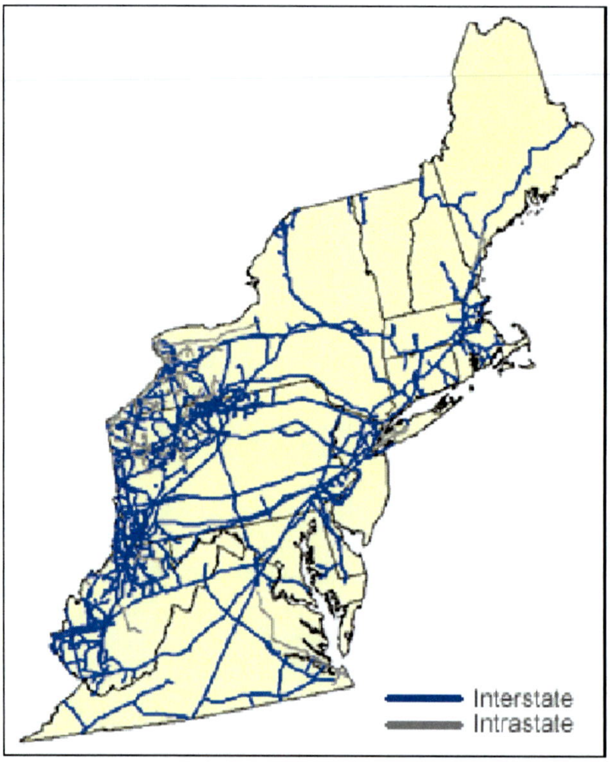

Source: Energy Information Administration, Office of Oil & Gas, Natural Gas Division, Gas Transportation Information System,' http://www.eia.doe.gov/pub/oil_gas/natural_gas/analysis_publications/ngpipeline/northeast.html.

Notes: Includes Connecticut, Delaware, Massachusetts, Maine, New Hampshire, New Jersey, New York, Pennsylvania, Rhode Island, Virginia, and West Virginia.

Figure 8. Northeast Region Natural Gas Pipeline Network

Groundwater Resource Issues

The U.S. Geological Survey identifies three principal hydrogeological environments overlying the Marcellus Shale: (1) glacial sand and gravel aquifers in New York, northern Pennsylvania, and northeastern Ohio, (2) valley-and-ridge carbonate rock and other aquifers in Pennsylvania and eastern West Virginia, and (3) Mississippian aquifers in northern Pennsylvania and northeastern Ohio. These aquifer systems are important supplies of fresh water for communities, especially in more rural areas, although in general most residents of these states obtain their drinking water from surface water sources. Typically, these aquifers are much closer to the ground surface than the Marcellus Shale, which can be thousands of feet deep. The groundwater wells in these states may reach only several hundred feet in depth.

The layers of rocks separating most fresh water aquifers from the Marcellus Shale are typically siltstones and shales layered with minor sandstones and limestone. Siltstones and shales generally act as barriers to fluid flow. These intervening layers of rocks can be several thousand feet thick in the eastern and northern portions of the area where the Marcellus Shale is deepest. On the western and southern portions of the area, the Marcellus Shale is shallower, as separated by a thinner package of siltstones and shales.

Many of the surficial sand and gravel aquifers form valley-fill deposits, in low-lying areas or stream valleys, and are recharged by precipitation that runs off surrounding, less permeable uplands.[36] As such, they would be particularly susceptible to leaky surface impoundments or careless surface disposal because of the relatively short distance and travel time from the land surface to the top of the water table. New York, for example, has deemed these unconsolidated sand and gravel aquifers "primary" or "principal" aquifers, which are highly productive and presently used as a significant source of water, or are a potentially abundant water supply.[37]

The EPA defines an aquifer that supplies at least 50 % of the drinking water consumed in the area overlying the aquifer as a "sole or principal source aquifer" (referred to, for convenience, as sole source aquifers).[38] Those who depend on the aquifer for drinking water may have no alternative drinking water source (physically, legally or economically). The EPA has designated at least four sole source aquifers in New York (EPA Region II) and one in Pennsylvania (Region III) that overlie the Marcellus Shale.[39] (See **Figure 9**). It is uncertain whether these areas would be targets for Marcellus Shale development.

DRILLING AND DEVELOPMENT TECHNOLOGY

In their early stages of development, conventional petroleum reservoirs depend on the pressure of their gas-cap and oil-dissolved gas to lift the oil to the surface (i.e., gas drive). Water trapping the petroleum from below also exerts an upward hydraulic pressure (water-drive). Petroleum reservoirs produced by the pressure of their natural gas and water drives are thus termed "conventional drive." As a reservoir's production declines, lifting further petroleum to the surface requires pumping—giving rise to the term "artificial lift." In the late 1940s, drilling companies began inducing hydraulic pressure in wells to fracture the producing formation. This stimulated further production by effectively increasing a well's contact with a formation. Advances in directional drilling technology now enable wells to deviate from nearly vertical to extend horizontally into the reservoir formation, which increases the well's contact with the reservoir. Directional drilling technology also enables drilling a number of wells from a single well pad, thus cutting costs while reducing environmental disturbance. Combining hydraulic fracturing with directional drilling has opened up production of tighter (less permeable) petroleum and natural gas reservoirs, and in particular, unconventional gas shales like the Marcellus.

Drilling

Well drilling has progressed from an art to a science. Originally, drillers used "cable-tool" rigs and a percussion bit. The drill operator would raise the bit and release it to pulverize the sediment. From time to time, the driller would stop and "muck out" the pulverized rock cuttings to advance the well. Though time-consuming, this method was simple and required minimal labor. Some drillers still use this method for water-wells and even some shallow gas-wells. The introduction of rotary-drill rigs at the beginning of the 20th century marked a big advance in drilling, particularly with the development of the "tri-cone rotary bit."[40] This method, as the name implies, uses a weighted rotating bit to penetrate the sediment (see **Figure 10**).

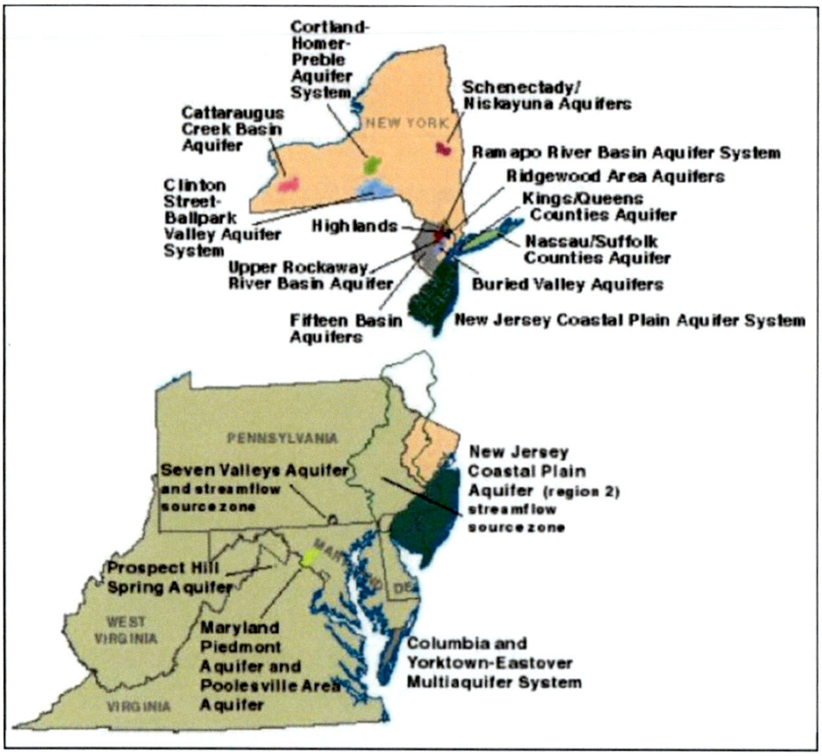

Source: U.S. EPA, *Sole Source Aquifer Protection Program*, Regions II and III, http://cfpub.epa.gov/safewater/ sourcewater/sourcewater.cfm?action=SSA (modified by CRS).

Notes: EPA defines a sole or principal source aquifer as an aquifer that supplies at least 50 % of the drinking water consumed in the area overlying the aquifer.

Figure 9. Sole Source Aquifers in the Appalachian—Northeast Region

The key to a rotary drill's speed is the relative ease of adding new sections of drill pipe (or drill string) while the drill-bit continues turning. Circulating fluids (drilling mud) down through the center of the hollow drill pipe and up through the well bore lifts the drill cuttings to the surface. Modern drill bits studded with industrial diamonds gives them an abrasive property to grind through any rock type. However, from time to time the drill string must be removed (a process termed "tripping") to replace the dulled drill bit.

To function properly, drilling fluids must lubricate the drill bit, keep the well bore from collapsing, and remove cuttings. The weight of the mud column prevents a "blow-out" from occurring when encountering high-pressure reservoir fluids. Drillers base the mud's composition on natural bentonite clay, a "thixotropic" material that is solid when still and fluid when disturbed. This essential rheological property keeps the drill cuttings suspended in the mud. The mud's chemistry and density must be carefully monitored and adjusted as the drilling deepens (for example, adding a barium compound increases mud density). "Mud pits," excavated adjacent to the drill rig provide a reservoir for mixing and holding the mud. The mud pits also serve as settling ponds for the cuttings. At the completion of drilling, the mud may be recycled at another drilling operation, but the cuttings will be disposed of in the pit. Several environmental concerns over drilling stem from the (hazardous) composition of the drilling mud and cuttings, and the potential for mud pits overflowing and contaminating surface water.

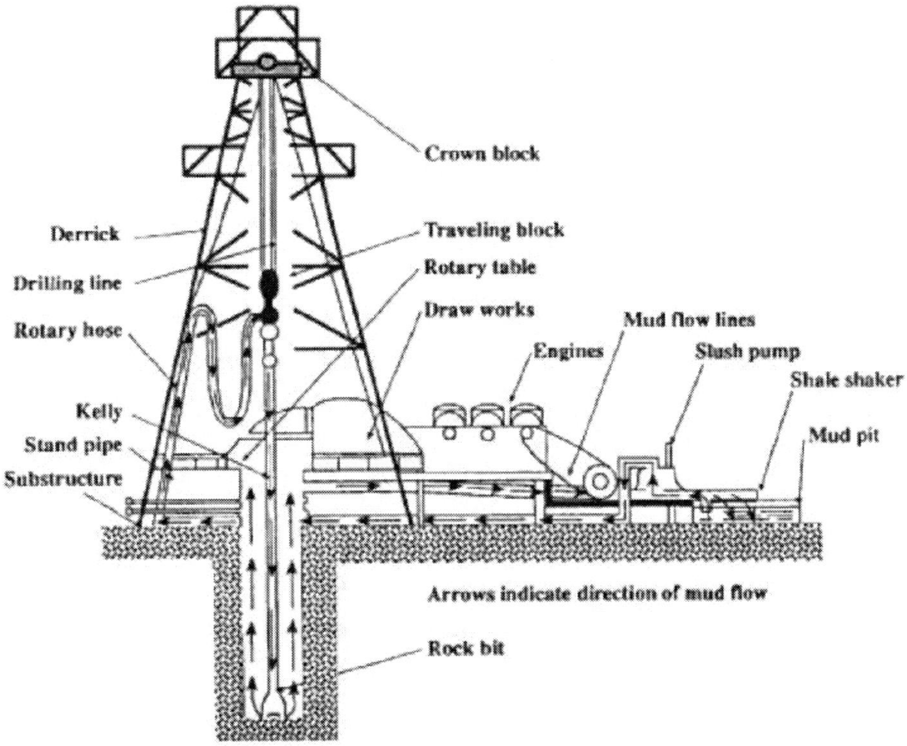

Source: Bureau of Land Management, http://www.blm.gov/rmp/wy/rawlins/ documents/ RMO/figures/Figure-A5.gif.

Figure 10. Rotary Drilling Rig

The most recent advance in drilling is the ability to direct the drill bit beyond the region immediately beneath the drill rig. Early directional drilling involved placing a steel wedge down-hole (whipstock) that deflected the drill toward the desired target, but lacked control and consumed time. Advances such as steerable down-hole drill motors that operated on the hydraulic pressure of the circulating drilling mud offered improved directional control. However to change drilling direction the operator had to halt drill string rotation in such a position that a bend in the motor pointed in the direction of the new trajectory (referred to as the sliding mode). Newer rotary steerable systems introduced in the 1990s eliminated the need to slide a steerable down- hole motor.[41] The newer tools drill directionally while continuously rotated from the surface by the drilling rig. This enables a much more complex, and thus accurate, drilling trajectory. Continuous rotation also leads to higher rates of penetration and fewer incidents of the drill-string sticking. (See **Figure 11**.)

Directional drilling offers another significant advantage in developing gas shales. In the case of thin or inclined shale formations, a long horizontal well increases the length of the well bore in the gas-bearing formation and therefore increases the surface area for gas to flow into the well. However, the increased well surface (length) is often insufficient without some means of artificially stimulating flow. In some sandstone and carbonate formations, injecting dilute acid dissolves the natural cement that binds sand grains thus increasing permeability. In tight formations like shale, inducing fractures can increase flow by orders of magnitude. However, before stimulation or for that matter production can take place, the well must be completed and cased.

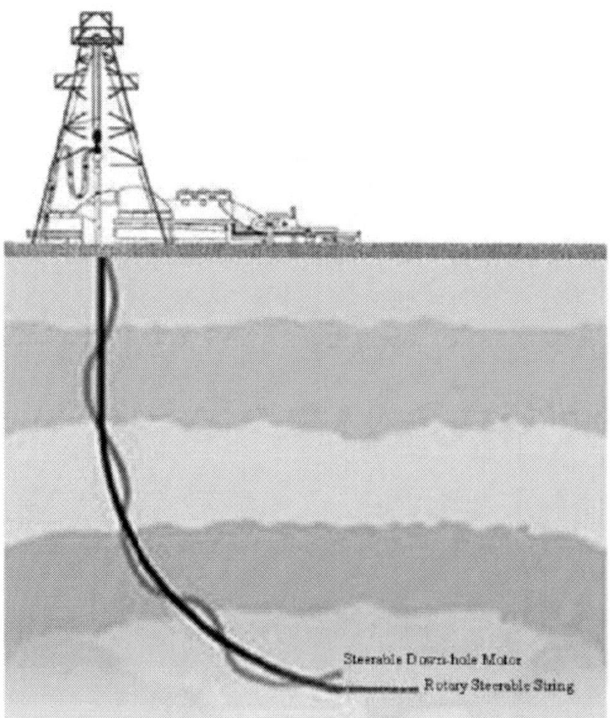

Source: Schlumberger, modified by CRS.

Figure 11. Directional Drilling Steerable Down-hole Motor vs. Rotary Steerable System

Well Construction and Casing

Commercial gas and oil, and municipal water-supply wells have in common a series of telescoping steel well casings that prevent well-bore collapse and water infiltration while drilling. The casing also conducts the produced reservoir fluids to the surface (see **Figure 12**). A properly designed and cemented casing also prevents reservoir fluids (gas or oil) from infiltrating the overlying groundwater aquifers.

During the first phase of drilling, termed "spuding-in," shallow casing installed underneath the drilling platform serves to reinforce the ground surface. Drilling continues to the bottom of the water table (or the potable aquifer), at which point the drill string is "tripped" out (removed) in order to lower a second casing string, which is cemented-in and plugged at the bottom. Drillers use special oil-well cement that expands when it sets to fill the void between the casing and the wellbore.

Surface casing and casing to the bottom of the water table prevents water from flooding the well while also protecting the groundwater from contamination by drilling fluids and reservoir fluids. (The initial drilling stages may use compressed air in place of drilling fluids to avoid contaminating the potable aquifer.) Drilling and casing then continue to the "pay zone"—the formation that produces gas or oil. The number and length of the casings, however, depends on the depth and the properties of the geologic strata.

After completing the well to the target depth and cementing-in the final casing, the drilling operator may hire an oil-well service company to run a "cement evaluation log." An electric probe lowered into the well, measures the cement thickness. The cement evaluation log provides the critical confirmation that the cement will function as designed—preventing well fluids from bypassing outside the casing and infiltrating overlying formations.

Absent any cement voids, the well is ready for completion. A perforating tool that uses explosive shape charges punctures the casing sidewall at the pay zone. The well may then start producing under its natural reservoir pressure or, as in the case of gas shales, may need stimulation treatment.

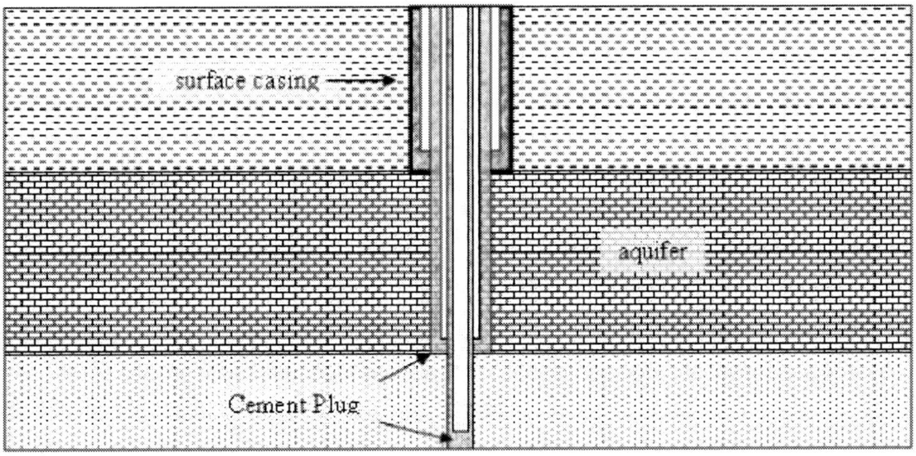

Source: CRS.
Notes: No Scale.

Figure 12. Hypothetical Well Casing

Both domestic-use gas wells and water wells are common throughout regions experiencing recent shale gas development. In the absence of regulation, domestic-use wells (gas or water) may not meet standard practices of construction. If the wellhead of a water supply well is improperly sealed, for example, surface water may infiltrate down along the casing exterior and contaminate the drinking-water aquifer. Some domestic water wells have also produced natural gas, and some shallow gas wells have leaked into nearby building foundations. To avoid some of these problems, Pennsylvania is instituting regulations that require a minimum 2,000-foot setback between a new gas-well and an existing water-well.

Hydraulic Fracturing

Despite some shales' abundant natural gas content, they do not produce gas freely. Economic production depends on some means of artificially stimulating shale to liberate gas. In the late 1940s, Texas oil fields responded to fluids pumped down wells under pressures high enough to fracture stimulated the producing formation. Hydraulic fracture stimulation treatments have been adapted to tight gas formations such as the Barnett Shale in Texas, and more recently the Marcellus Shale.

Typical "frac" treatments or frac jobs (as commonly referred to) are relatively large operations compared to some drilling operations. The oilfield service company contracted for the work may take a week to stage the job, and a convoy of trucks to deliver the equipment and materials needed (see **Figure 13**).

A company involved in developing Texas gas shale offered the following description of a frac job:[42]

> Shale gas wells are not hard to drill, but they are difficult to complete. In almost every case, the rock around the wellbore must be hydraulically fractured before the well can produce significant amounts of gas. Fracturing involves isolating sections of the well in the producing zone, then pumping fluids and proppant (grains of sand or other material used to hold the cracks open) down the wellbore through perforations in the casing and out into the shale.
>
> The pumped fluid, under pressures up to 8,000 psi, is enough to crack shale as much as 3,000 ft in each direction from the wellbore. In the deeper high-pressure shales, operators pump slickwater (a low-viscosity water-based fluid) and proppant. Nitrogen-foamed fracturing fluids are commonly pumped on shallower shales and shales with low reservoir pressures.

Fracturing Fluids

Fracturing fluid functions in two ways: opening the fracture and transporting the "propping" agent (or proppant) the length of the fracture.[43] As the term propping implies, the agent functions to prop or hold the fracture open. The fluid must have the proper viscosity and low friction pressure when pumped, it must breakdown and cleanup rapidly when treatment is over, and it must provide good fluid-loss control (not dissipate). The fluid chemistry may be water-based, oil- based or acid-based depending on the properties of the formation. Water-based fluids (sometimes referred to as slickwater) are the most widely used (especially in shale formations) because of their low cost, high performance, and ease of handling. Some fluids may also include nitrogen and carbon dioxide to help foaming. Oil-based fluids find use in hydrocarbon bearing formations susceptible to water damage, but are they expensive and difficult to use. Acid-based fluids use hydrochloric acid to dissolve the mineral matrix of carbonate formations (limestone and dolomite) and thus improve porosity; the reaction produces inert calcium chloride salt and carbon dioxide gas.

Water-based fluids consist of 99% water with the remainder made up of additives. The initial fracturing stage may use hydrochloric acid (HCl) to cleanup the wellbore damage done during drilling and cementing. Gelling agents, based on water-soluble polymers such as vegetable- derived guar gum, adjust frac fluid viscosity. The most widely used additives for breaking down fluid viscosity after fracturing are oxidizers such as ammonium (NH^{+4}), potassium, and sodium salt of peroxydisulfate ($S2O_8^{-2}$); enzyme breakers may be based on hemicellulase (actually a mixture of enzymes which can hydrolyze the indigestible components of plant fibers). Silica flour serves as good fluid-loss additive. Biocides added to polymer-containing fluids prevent bacterial degradation (as the polysaccharides (sugar polymer) used to thicken water are an excellent food source for bacteria). Methanol (an alcohol) and sodium thiosulfate ($Na_2S_2O_3$) are commonly used stabilizers added to prevent polysaccharide gels degrading above temperatures of 200°F.

It is important to note that the service companies adjust the proportion of frac fluid additives to the unique conditions of each well. The Occupational Safety and Hazard Administration (OSHA) requires that material safety data sheets (MSDS) accompany each chemical used on the drill site, but the proportion of each chemical additive may be kept proprietary.[44]

Proppants hold the fracture walls apart to create conductive paths for the natural gas to reach the wellbore. Silica sands are the most commonly used proppants. Resin coating the sand grains improves their strength.

Source: Chesapeake Energy Corporation, 2008.

Notes: The yellow frac tanks in the foreground and along the tree line hold water, the red tanker holds proppant; hydraulic pumps are in the center.

Figure 13. Hydraulic Fracture Job Marcellus Shale Well, West Virginia

Hydraulic Fracture Process

Fracture treatments are carefully controlled and monitored operations that proceed in stages. Before beginning a treatment, the service company will perform a series of tests on the well to determine if it is competent to hold up to the hydraulic pressures generated by the fracture pumps.

In the initial stage, an HCl solution pumped down the well cleans up residue left from cementing the well casing. Each successive stage pumps discrete volumes of fluid (slickwater) and proppant down the well to open and propagate the fracture further into the formation. The treatment may last upwards of an hour or more, with the final stage designed to flush the well. Some wells may receive several or more treatments to produce multiple fractures at different depths, or further out into the formation in the case of horizontal wells.

A single fracture treatment may consume more than 500,000 gallons of water.[45] Wells subject to multiple treatments consume several million gallons of water. An Olympic-size swimming pool (164 ft x 82 ft x 6 ft deep) holds over 660,000 gallons of water, for comparison, and the average daily per capita consumption of fresh water (roughly 1,430 gallons per day) is 522,000 gallons over one year.[46]

The high injection pressure not only opens and propagates the fracture but also drives fluid into the shale's pore spaces. A high volume of fluid also remains in the fracture that will impede gas flow to the well if not pumped out. The subsequent "flowback" treatment performed, attempts to recover as much of the remaining fluid as possible without removing the proppants. The "flowback" water pumped out of the well may be high in dissolved salts and frac chemicals, however, making it unsuitable for beneficial use, and requiring treatment for disposal. After the well begins producing gas, it may also produce more flowback water. Flowback disposal presents environmental issues, as discussed in the "Surface Water Quality Protection" section below.

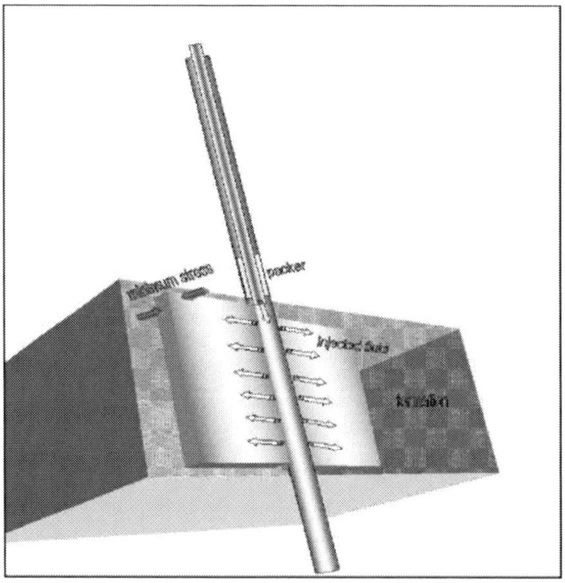

Source: CRS.

Notes: Fracture opens perpendicular to direction of minimum stress.

Figure 14. Idealized Hydraulic Fracture

Fracture Geometry

The fracture is ideally represented by a vertical plane that intersects the well casing (**Figure 14**). It does not propagate in a random direction, but opens perpendicular to the direction of least stress underground (which is nearly horizontal in orientation).

Hydraulic fracturing attempted at too shallow a depth develops an undesirable horizontal fracture. In the case of deep shales, such as the Marcellus, this appears unlikely. The fracture may extend outward several thousands of feet, but the fracture height is limited to the thickness of the shale formation (and controlled by the perforated zone of the well).

Fracturing Risks to Groundwater

The geologic environment that led to the deposition of the Marcellus Shale, and the overlying layers of siltstone, shale, sandstone, and limestone have kept gas from the Marcellus Shale confined at depth, and prevented it from naturally migrating upwards toward the surface. The process of developing a shale gas well (drilling through an overlying aquifer, stimulating the well via hydraulic fracturing, completing the well and producing the gas) is an issue of concern for increasing the risk of groundwater contamination. Typically, well drilling and completion practices, as described above, require sealing the well by casing throughout the aquifer interval. A properly designed and cased well will prevent drilling fluids, hydraulic fracturing fluids, or natural gas from leaking into the permeable aquifer and contaminating groundwater. The casing also prevents groundwater from leaking into the well where it could interfere with the gas production process.

An environmental concern raised by hydraulic fracturing is the possibility of introducing contaminants into aquifers. Hydraulic fracturing does induce new fractures into shale, and can propagate fractures thousands of feet along the bedding plane of a shale formation. The potential for propagating fractures to an overlying aquifer may depend on the depth separating the two. However, engineers designing and conducting frac jobs have a strong incentive to limit the fractures to the height of the gas-producing shale zones. Any fracture propagated to an overlying aquifer could allow water to flow down into the gas-producing portion of the shale, which could significantly hamper gas production.

Multiple frac treatments may pump as much as 2 to 3 million gallons of fluid down a well. Afterwards the well operator recovers a large proportion of these fluids by pumping them out of the well, and disposes of them through waster-water treatment plants or by other means as discussed below. However, any drilling fluids or frac fluids spilled on the ground surface could infiltrate downwards to shallow groundwater. This could pose a risk to surficial aquifers composed of very permeable unconsolidated sand and gravel deposits, such as those in northern Pennsylvania and southern New York (see **Figure 9**).

Another potential issue could be groundwater contamination from poorly constructed water wells. Generally, drinking water wells are shallower than natural gas wells, and their casing may not extend their entire depth. This is particularly the case for domestic water wells that may not be subject to the same level of oversight and scrutiny as natural gas well construction. A water well that is not cased from the surface, or is not constructed and cased properly, might allow contaminated water to flow from the ground surface and enter the water well, possibly compromising the quality of drinking water in the well and even the drinking water aquifer itself. In such instances, and particularly where natural gas drilling and stimulation activities are nearby, leaky surface impoundments or careless surface disposal of drilling fluids at the natural gas operation could increase the risk of contaminating the nearby water well.

Contaminated surface water unrelated to drilling could also contaminate improperly cased or constructed water well. For example, a leaky septic system, or improper disposal of domestic refuse such as car batteries or used oil, could leak from the surface into the water well. If this is the case, a dispute could ensue as to who is responsible for contaminating the water well. Resolving the dispute could involve a hydrogeological investigation to prove or disprove any linkage between natural gas development activities and water well contamination, often at considerable expense and with an uncertain outcome, given the complexity of groundwater flow at most sites.

LEASING ISSUES FOR GAS DEVELOPMENT

The Marcellus Shale gas formation has generated considerable interest within the past two years on state-owned and private lands throughout the Appalachian region in West Virginia, Pennsylvania, and New York. Landowners who also own the subsurface mineral rights may lease their land for the development of those minerals or sell the surface or mineral rights to qualified buyers for development.[47] If the landowner decides to lease the land for mineral development, the owner may negotiate a lease agreement or contract. Common features of oil and gas leases today include signing bonuses (or up front "bonus bid" payments for competitive lease sales), royalties, rents, primary lease terms and conditions for lease renewal.

In the case of mineral rights severed from surface rights, a split-estate is established. Under a split-estate, the mineral owner has the right to reasonable access to recover the mineral. The surface owner typically has the right to protection from "unreasonable encroachment and damage" and can usually negotiate compensation for the use of the surface and for damages. On split-estates that involve private surface owners and federally owned minerals, a federal lessee must meet one of the following conditions: a surface agreement; written consent or a waiver from the private surface owner for access to the leased lands; payment for loss or damages; or the execution of a bond not less than $1,000.[48] Split-estate scenarios are common but a comprehensive discussion on the subject is beyond the scope of this chapter.

The state-by-state sections below describe how compensation to landowners atop the Marcellus Shale has evolved over the past few years. The discussion focuses primarily on bonus payments and royalties received by state and private landowners today compared to what they received just a few years ago. Rents are not part of the discussion because nearly all of the information available reports on signing bonuses and royalties. Further, lease agreements often roll signing bonuses into rents, and call for rents paid upfront or paid quarterly as a "delay rental." Rents appear to be much less significant to small landowners who lease a few acres. On state and private lands, as with federal lands, lessees usually pay rent until production commences and then switch to paying royalties on the value of production.

New York

In 1999, signing bonuses for leases on private lands were around $5 per acre, with royalties of 12½ %. During that same time, the state received between $15 and $600 per acre in bonus payments. In today's climate, bonus payments are as much as $3,000 per acre on privately held lands with royalty rates between 15% and 20%. Lease terms are typically three to five years with renewal clauses for continued production and shut-in wells. Fewer than 10 years ago, there were only about five or six companies involved in securing leases to develop gas shale projects. By 2008 their number rose to about four dozen companies (including Exxon/Mobil, BP, and Conoco Phillips).

Pennsylvania

In 2002, companies were paying signing bonuses of $25 to $50 per acre and agreeing to $2 per acre annual rentals and 121/2 % royalty on 10-year leases on state-owned land. Interest has grown dramatically since that time, and firms are now paying as much as $2,500 per acre in bonus payments, 16% royalty and annual rental rates of $20 per acre for years 2 through 5 and $35 per acre in years 6 through 10. Private landowners have typically issued shorter-term leases of 5 to 7 years. Private owners received about $12 per acre in signing bonuses and a 121/2% royalty rate in 2003. In 2008, lessees are now paying private landowners signing bonuses of nearly $2,900 per acre with 17% to 18% royalty rates on 5- through 7-year leases. Many private leases often calculate rents as part of the signing bonus and sometimes pay them upfront or over time.

West Virginia

Some landowners in West Virginia have seen their bonus bids climb from $5 per acre in 2007 and early 2008, to more recent bonus payments of $1,000 to $3,000 per acre. Royalty rates have increased from 121/2% through 16% to 18%. Rents are often included in the signing bonus or sometimes paid out in the form of a "delay rental."

State Summary

In general, Marcellus gas lease rates range for state and private lands from a low of 121/2% in West Virginia to high of 20% in New York (**Table 1** and **Table 2**), but these rates are less than those paid in Texas for similar gas shale tracts. All three of Marcellus Shale states reported above have shown significant increases in the amounts paid as signing bonuses and increases in royalty rates. There are lease sales that the Natural Gas Leasing Tracking Service reported with signing bonuses as low as $100 to $200 per acre, owing to greater uncertainty and less interest among natural gas companies and/or the lack of information among landowners on what the land is worth.[49]

Table 1. Shale Gas Bonus Bids, Rents, and Royalty Rates on Selected State Lands

	Statutory Minimum or Standard Royal Rate	Royalty Rate Range	Bonus Bids (per acre)	Comments
West Virginia [a]	12½%	n.a.	n.a.	No state shale gas leases.
Pennsylvania [b]	12½%	12½% - 16%	$2,500	In many cases bonus bids were in the $25-$50 per acre range as recent as 2002. Aroyalty rate of 12.5% was most common.
New York [c]	12½%	15% - 20%	about $600	Bonus bids ranged from $15-$600 per acre around 1999-2000 and most royalty rates were at 12.5%.
Texas	12½%	25%	$350 - $400 (Delaware Basin) $12,000 (river tracts)	Bonus bids have been relatively consistent in recent times (within the past 5 years). Royalty rates were more common at 20%-25% about 5 years ago. Most state-owned lands are not considered to be among the best sites for shale gas development.

Source: CRS.

a. Personal communication with Joe Scarberry in the WV Department of Natural Resources, October 2008.

b. Personal communication with Ted Borawski in the PA Bureau of Forestry, who provided information on shale gas leases on both state and private lands, October 2008.

c. Personal communication with Lindsey Wickham of the NY Farm Bureau and Bert Chetuway of Cornell University, discussed lease sales on state and private land, October 2008.

Table 2. Shale Gas Bonus Bids, Rents, and Royalty Rates on Private Land in Selected States

	Royalty Rates Range	Bonus Bids (per acre)	Comments
West Virginia[a]	12 ½ % –18 %	$1,000 – $3,000	Bonus payments were in the $5 per acre range as recently as 1-2 years ago. Royalty rates were 12 ½ %
Pennsylvania	17% – 18%	$2,000 – $3,000	
New York	15% – 20%	$2,000 – $3,000	
Texas	25% – 28%	$10,000 – $20,000	Bonus bids were in the $1,000 range around 2000-2001. Royalty rates ranged from 20% to 25% range.

Source: CRS

a. Personal communication with David McMahon, Director of the WV Surface Owners Rights Organization, October 2008.

There are several landowner organizations that have formed in recent years to pool their land for leasing, advise landowners, and serve as information centers. Some groups seek out competitive bids from energy companies.[50]

Lease Audit (Product Valuation and Verification)

Organizations involved with private-owned leases for shale gas development often recommend to lessors that they include provisions/clauses in their leases that require the producer to pay the wellhead price without deductions or to base the royalty on the gross proceeds at the well. Under most circumstances (in arms-length, third-party transactions), the wellhead is the point of first sale. Wellhead and spot prices are available from different sources including the EIA website. There are opportunities for a lessor to audit volume flows from the well, but more often, the producer will provide production and price data to the lessor. According to David McMahon of the West Virginia Surface Owners Rights Organization (WVSORO), there are major concerns among landowners that include the lessees deducting costs inappropriately, before the royalty is calculated.[51] This practice can deduct costs for transportation, compressors, and line loss from the wellhead price, thus reducing the royalty paid to the landowner. While metering production is typical, auditing production may be rare. The cost of an audit may be cost prohibitive for small landowners.

The state of Pennsylvania audits production from the top 100 wells of its natural gas leases using an independent auditor, or "meter truck companies" that work for the shale gas producers. The state of New York requires that natural gas producers meter production and make that information available upon request.[52] The West Virginia Department of Environmental Protection, Office of Oil and Gas, requires an annual production report from all oil and gas producers in the state.

Severance Taxes

State governments of Pennsylvania and New York have proposed a severance tax (included in the 2009-20 10 budget) in order to generate revenue to address state concerns. Twenty-seven states producing natural gas (including West Virginia) already impose a severance tax. In Pennsylvania, based on the proposed severance tax, the Pennsylvania state government projected revenue increases of $107 million in 2009-2010, increasing to $632 million by 2013-2014.[53] A separate study on the economic impacts of Marcellus Shale gas development in Pennsylvania projected economic output of $3.8 billion, 48,000 jobs, and $400 million in state and local tax revenues in 2009.[54] The study also projected cumulative tax revenues reaching an estimated $12 billion (in present value dollars) in 2020. However, based on the proposed severance tax, state and local tax revenue estimates could be reduced by $880 million (in present value dollars) resulting from an estimated 30% reduction in natural gas drilling between 2009 and 2020 according to the study.

There are typically two types of severance taxes: an ad valorem or value-based tax or a unit of production-based tax. Individual states uniquely apply severance taxes to the extraction industries for the "privilege" of extracting natural resources (such as minerals and timber) in order to compensate residents for the irreplaceable loss from the extraction of the resource, sometimes described as a loss of future wealth. Revenue streams from severance taxes pay for the environmental and social costs associated with mining and extraction. The tax may also contribute to a state's general revenue fund.

The severance tax as other costs imposed on the oil and gas industry may affect long-term production rates. This variable cost, applied to the operator, only when production occurs, may slow extraction. A whole host of other factors, e.g., geology, price/demand, transportation costs, and capital formation, affect when and where natural gas is developed.

Federal Land Leasing and Restrictions to Leasing

Leasing federal public lands for oil and gas development is based on multiple-use/sustained yield Resource Management Plans (RMPs) prepared by the Bureau of Land Management (BLM) in the Department of the Interior. In accordance with those management plans, BLM offers tracts of public land with oil and gas potential for competitive leasing each quarter. The BLM administers oil and gas leasing and development on federally owned minerals both for BLM lands and on behalf of the U.S. Forest Service, which in turn administers the surface development within the National Forests and through its land planning process addresses surface occupancy concerns and approves or rejects any extraction of forest resources.

However, privately held mineral rights account for over 90% of the Allegheny National Forest in Pennsylvania and about 38% within the Monongahela National Forest in West Virginia. Finger Lakes National Forest in New York was withdrawn from the mineral leasing programs under section 370 the Energy Policy Act of 2005 (P.L. 109-58).

The Mineral Leasing Act and the Outer Continental Shelf Lands Act, as amended, requires that all public lands (onshore and offshore areas respectively) available for lease be offered initially to the highest responsible qualified bidder by competitive bidding. The objective of the competitive bid is to provide a "fair market value" return to the federal government for its resources. Under the two acts mentioned above, the federal government requires royalties from oil and gas producers (and other resource producers) on leasable federal lands and annual rents from non- producing lessees. Competitive oil and gas lease sales include bonus bids (upfront payments made to obtain a lease). The statutory minimum royalty rate for oil and gas leases on public lands is 12½ % (also expressed as 1/8). Currently the royalty rate for onshore oil and gas leases is 12½ %, while the rate for an offshore lease can range from 12 ½ % to 18¾ %. Annual rental rates established for onshore leases are at not less than $1.50 per acre for the first five years of a ten- year lease and not less than $2 per acre each year thereafter. The primary lease term for an onshore lease is 10 years and for an offshore lease is either 5, 8, or 10 years depending on the depth of the water. Annual rent paid on leases ends when production begins, then the lessor begins paying royalties on the value of production.

Of the approximately 5 million acres of federal land in the Appalachian basin (**Figure 15**), 46% (2.5 million acres) is not accessible for lease.[55] Based on resource estimates, these lands contain 41% (984.7 bcf) of the federal natural gas in the basin. Another 42% (2.2 million acres) is accessible with restrictions on oil and gas operations beyond standard lease terms. These lands contain 45% (1.1 tcf) of the federal natural gas. The remaining 13% (691.7 thousand acres) that is accessible under standard lease terms contains 14% (346.7 bcf) of the federal natural gas. Most of the undiscovered gas resource (94%) will likely occur in continuous accumulations. Coal-bed natural gas accounts for about 13% of the total undiscovered continuous gas.

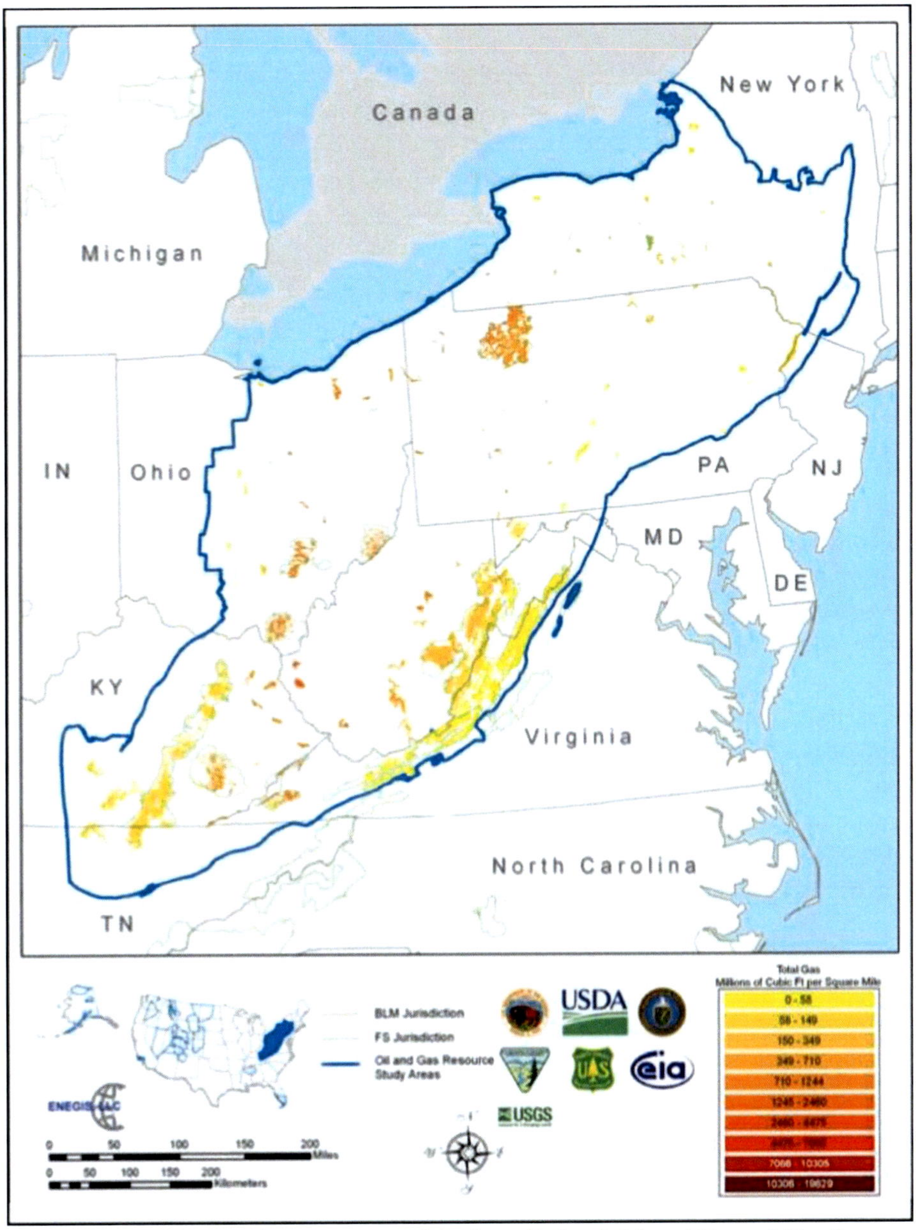

Source: U.S. Dept. of the Interior, *Inventory of Onshore Federal Oil and Natural Gas Resources and Restrictions to Their Development, Phase III Inventory*, 2008.

Notes: Approximately 5 million acres of federal lands in the Appalachian Basin region overlie natural gas resources.

Figure 15. Federal Lands Overlying Natural Gas Resources of the Appalachian Basin

FEDERAL AND STATE LAWS AND REGULATIONS AFFECTING GAS SHALE DEVELOPMENT

Development of the Marcellus Shale will be subject to regulation under several federal and state laws. In particular, the large volumes of water needed to drill and hydraulically fracture the shale, and the disposal of this water and other wastewater associated with gas extraction may pose significant water quality and quantity challenges that trigger regulatory attention. As the U.S. Geological Survey noted in a recent publication, "Concerns about the availability of water supplies needed for gas production, and questions about wastewater disposal have been raised by water-resource agencies and citizens through the Marcellus Shale gas development region."[56] The following sections review key provisions of two relevant federal laws, the Safe Drinking Water Act (SDWA) and the Clean Water Act (CWA), and related state requirements.

Surface Water Quality Protection

As previously described, hydraulic fracturing involves injecting water, sand, and chemicals into the shale layer at extremely high pressures in order to release the trapped natural gas. It is a water-intensive practice. Typical projects use 1-3 million gallons of water for each well and 0.5 million pounds of sand. Large projects may require up to 5 million gallons of water.

The Texas Railroad Commission estimates that fracturing a vertical well in the Barnett Shale in Texas can use more than 1.2 million gallons of water, while fracturing a horizontal well can use more than 3.5 million gallons.[57] Moreover, the wells may be re-fractured several times, thus requiring additional water. Fracturing operations use an estimated 5 or more million gallons of water each day in the Barnett Shale, a smaller natural gas field in Texas.[58] Regarding the Marcellus Shale region, the USGS observed "many regional and local water management agencies are concerned about where such large volumes of water will be obtained, and what the possible consequences might be for local water supplies."[59]

Some of the injected fluids remain trapped underground, but the majority of the injected water— 60% to 80%—returns to the surface as "flowback" after the frac treatment. It typically contains proppant (sand), chemical residue, and trace amounts of radioactive elements that naturally occur in many geologic formations.[60] USGS notes that because the quantity of fluid used is so large, the additives in a three million gallon frac job would yield about 15,000 gallons of chemicals in the waste.[61] The well service company may temporarily retain the flowback and brine in open-air, lined retention ponds before reusing it (if possible), or disposing it. The drilling operator must reclaim the temporary storage pits when the drilling and fracturing operations end. In addition, the well operator must separate, treat, and dispose the natural brine co-produced with gas. As noted below, where feasible, the produced water may be disposed through underground injection. The oil and gas industry uses this disposal method in some western states and in Ohio.[62] The industry has not yet begun using it as a disposal alternative for gas production in eastern Marcellus Shale.

In the event that underground injection is not feasible in the area of the Marcellus Shale, the well service company may discharge the flowback to surface waters if the discharge does not violate a stream or lake's water quality standards. Standards established by states under Section 303 of the Clean Water Act (CWA) protect designated beneficial uses of surface waters, such as recreation or public water supply.[63]

If contaminants present in the flowback prevent discharge to surface water without further treatment, it is likely that the service company will have to transfer the wastewater off-site to an industrial treatment facility or a municipal sewage treatment plant that is capable of handling and processing the wastewater. In this case, the operator of the publicly owned treatment works (POTW) or industrial treatment facility would assume responsibility for treating the waste before discharging it into nearby receiving water in compliance with effluent limits contained in the facility's discharge permit.[64] The chemical frac additives returned in flowback and the produced brine could cause problems for POTWs. Contaminants in industrial process wastewaters can kill off the biota essential to a POTW's operation. If contaminants pass through the POTW without adequate treatment, the discharge could violate the facility's discharge permit and could cause a violation of water quality standards. A standard POTW's effective treatment of flowback and brine is uncertain. It could require upgrading, but the cost of such an upgrade is also uncertain.

In the fall of 2008, water samples from the mid-Monongahela River valley of Pennsylvania showed high levels of total dissolved solids (TDS), which indicate salinity. Although the TDS was determined to pose little threat to health or safety, preliminary analysis suggested that the principal source was large truck deliveries of wastewater from gas well drilling sites in the Marcellus Shale to POTWs discharging, directly or indirectly, into the Monongahela River. In October 2008, state officials ordered nine sewage treatment plants to reduce their volumes of gas well drilling water, which contains high concentrations of TDS. Subsequent analysis concluded that discharge from abandoned mines was more responsible for the high TDS than drilling wastewater discharges from municipal wastewater treatment plants.[65] However, state officials remain concerned about the projected need for treatment of wastewater (both initial flowback water from fracturing and longer term production brines) from gas well development—projected to be as much as 20 million gallons per day in 2011—and the capacity of the state's surface waters to assimilate associated wastewaters. In April 2009, the Pennsylvania Department of Environmental Protection issued a permitting strategy document for gas development wastewaters, requiring that any discharges will be subject to the most stringent treatment or water quality standards needed to protect aquatic life in the state's streams. Their goal is prohibiting new sources of high-TDS wastewaters from discharging into Pennsylvania's waters by January 1, 2011.[66]

Brine storage and transport are major issues in developing the Barnett Shale in Texas, and are likely to be key issues in development of the Marcellus Shale, as well. Currently, permitted treatment facilities capable of treating such wastes are not adequate. In Pennsylvania, there are five facilities designed to treat the type of industrial wastewater that is involved in producing shale gas. Most of the well sites are located in northeast Pennsylvania, while the closest treatment facilities are nearly 250 miles away.[67]

West Virginia, too, recognizes that wastewater disposal is "perhaps the greatest challenge regarding these operations."[68] State officials say that underground injection control (see discussion below) may be the best option for wastewater disposal, but the state has only permitted two commercial underground injection control (UIC) wells. The state has no centralized commercial treatment facilities available, and state officials are cautious about the capability of POTWs to handle the flow and quality of waste that they might receive.[69] The

West Virginia Department of Environmental Protection has proposed both changes to the state's oil and gas drilling rules (which the state legislature must approve) and an industry guidance document to assist operators in planning for the water issues associated with drilling and operating these wells. However, local groups have criticized the proposed rules and draft non-binding guidance for failing to address disposal of wastewater, disclosure of chemicals used in hydraulic fracturing, and where the additional quantities of water required for drilling will come from.[70]

One potential solution to off-site disposal may be on-site treatment and reuse; that is, treating and reusing flowback and produced water on-site. Some companies are reportedly considering on-site treatment options such as advanced oxidation and membrane filtration processes. On-site treatment technologies may be capable of recovering 70%-80% of the initial water to potable water standards, thus making the water immediately available for reuse. The remaining 20%-30% is very brackish and considered brine water. A portion may be further recoverable as process water, but not to achieve potable water standards. In other cases, companies send the brine water off-site for treatment and disposal. The economics of any such options are critical, and site factors such as available power and final water quality are often the determinant in treatment selection.

Other Surface Water Quality Issues

Another potential source of water pollution from oil and gas drilling sites is runoff that occurs after a rainstorm. The storm water runoff can transport sediment to nearby surface water bodies. Provisions of the CWA generally regulate storm water discharges from industrial and municipal facilities by requiring implementation of pollution prevention plans and, in some cases, remediation or treatment of runoff.[71] Industries that manufacture, process, or store raw materials and that collect or convey storm water associated with those activities are subject to the act's requirements. Furthermore, fracturing fluid chemicals and wastewater can leak or spill from injection wells, flow lines, trucks, tanks or holding pits and thus may contaminate soil, air, and water resources.

However, the act specifically exempts the oil and gas industry from these storm water management regulatory provisions. CWA Section 402(l)(2) exempts mining operations or oil and gas exploration, production, processing, or treatment operations or transmission facilities from federal storm water regulations, and Section 502(24) extends the exemption to construction activities, as well.[72] Thus, federal law contains no requirements to minimize uncontaminated sediment pollution from the construction or operation of oil and gas operations. However, the federal exemption does not hinder states from requiring erosion and sedimentation controls at well sites, under authority of non-federal law. Pennsylvania, for example, requires well drill operators to obtain a permit for implementation of erosion and sedimentation controls, including storm water management, if the site disturbance area is more than five acres in size. If the site is less than five acres, a plan for erosion and sediment control is required. Storm water requirements are part of this permit.[73] New York has similar requirements for erosion and sedimentation controls at well sites, regardless of site area. The Delaware River Basin Commission, which has jurisdiction over water quality in a portion of the area underlain by the Marcellus Shale (see section on State Water Quality Laws below) also has similar requirements regardless of site area.

Groundwater Quality Protection

Because development of the Marcellus Shale is dependent on the use of hydraulic fracturing, some fear it could potentially contaminate underground aquifers that provide water supplies to private wells and public water systems. The Safe Drinking Water Act in 2005 broadly excluded the underground injection of fluids used in hydraulic fracturing for oil and gas development.[74] However, the SDWA does not preempt states from imposing their own laws and regulations, and the states have long been responsible for protecting groundwater resources during oil and gas production activities.[75] For example, in New York, the Department of Environmental Conservation (DEC) has authority over oil and gas development in the state, including oversight of hydraulic fracturing activities to ensure protection of groundwater resources. Although federal laws do not regulate the injection of hydraulic fracturing fluids, states such as Pennsylvania and New York do require submission of information on hydraulic fracturing fluid composition prior to issuing a well permit. Moreover, other injection wells related to oil and gas development may be subject to federal requirements.

Safe Drinking Water Act Authority

The underground injection control provisions of the SDWA require the Environmental Protection Agency (EPA) to regulate the underground injection of fluids (including solids, liquids, and gases) to protect underground sources of drinking water. UIC program regulations specify sitting, construction, operation, closure, financial responsibility, and other requirements for owners and operators of injection wells.[76]

West Virginia, Ohio, and Texas are among the states that have assumed primacy and have lead implementation and enforcement authority for the UIC program, including primacy for injection wells associated with oil and gas development. EPA implements the programs directly for New York and Pennsylvania.[77] Most states, including Ohio and West Virginia, have received primacy for Class II oil and gas wells under Section 1425.

The Safe Drinking Water Act specifies that the UIC regulations may not interfere with the underground injection of brine or other fluids brought to the surface in connection with oil and gas production or any underground injection for the secondary or tertiary recovery of oil or natural gas

> unless such requirements are essential to assure that underground sources of drinking water will not be endangered by such injections.[78]

Additionally, the Energy Policy Act of 2005 amended SDWA UIC provisions to specify further that the definition of "underground injection" excludes the injection of fluids or propping agents (other than diesel fuels) used in hydraulic fracturing operations related to oil, gas, or geothermal production activities.[79]

The key statutory provisions are:[80]

- SDWA Section 1421 directs EPA to promulgate regulations for state underground injection control (UIC) programs, and mandates that the regulations contain minimum requirements for programs to prevent underground injection that endangers drinking water sources.[81]
- Section 1422 authorizes EPA to delegate primary enforcement authority (primacy) for UIC programs to the states, provided that state programs prohibit any underground injection that is not authorized by a state permit.[82]
- Section 1425 provides separate authority for states to attain primacy specifically for oil and gas (i.e., Class II) wells. The provision allows states to demonstrate that their existing programs for oil and gas wells are effective in preventing endangerment of underground sources of drinking water, providing an alternative to meeting many of the detailed requirements promulgated to implement the UIC program under Section 1421.
- Section 1431 grants the EPA Administrator emergency powers to issue orders and commence civil action to protect public water systems or underground sources of drinking water. The Administrator may take action when (1) a contaminant present in or likely to enter a public drinking water supply system or underground drinking water source poses a substantial threat to public health, and (2) state or local officials have not taken adequate action.[83]

Underground Injection of Waste Fluids

As noted, nearly all of the water injected for hydraulic fracturing must come back out of the well for gas to flow out of the shale. A key issue related to Marcellus Shale gas production is safely disposing large quantities of potentially contaminated fluids recovered from the gas wells.

EPA generally categorizes injection wells associated with oil and gas production as Class II wells under its UIC regulatory program. These are wells used to inject brines and other waste fluids associated with oil and natural gas production.[84] Given the expense of treating and transporting large volumes of wastewater for disposal, it is possible that the production of gas from the Marcellus Shale will increasingly involve the use of injection wells to dispose of poor-quality formation water, flowback water resulting from hydraulic fracturing, and other waste fluids associated with gas production.

EPA reports that most of the fluid injected into Class II wells has been brine brought to the surface in producing oil and gas. This brine, a naturally occurring formation fluid, is often very saline and may contain toxic metals and naturally occurring radioactive substances. According to EPA, the brine "can be very damaging to the environment and public health if it is discharged to surface water or the land surface."[85] To prevent contamination of land and surface water, Class II wells provide a means for disposing brines by re-injecting them back into their source formation or into similar formations. Injection wells also serve as disposal means for residual water from drilling and hydraulic fracturing operations. As states have adopted rules to prevent the disposal of saline water to surface water and soil, injection has become the preferred way to dispose of this waste fluid, where the local geology permits.[86]

In New York and Pennsylvania, both EPA and the state environmental agency must issue permits if the disposal method for fracturing wastewater is deep well injection. Pennsylvania law provides that "a well operator who affects a public or private water supply by pollution or diminution shall restore or replace the affected supply with an alternate source of water adequate in quantity and quality for the purposes served by the supply."[87] Additionally, it requires a permit application for a disposal well or enhanced recovery well to include an erosion and sedimentation plan for the well site.[88]

As of now, it is unknown how much water the gas wells in the Marcellus Shale formation will produce. The amount of water produced could vary across the region. Because shale gas formations generally are impermeable, they typically produce much less water than traditional oil and gas fields or coalfields. The impermeability of the shale also indicates that reinjection of wastewater from fracturing into the shale formation may not be feasible in many locations. Consequently, it is uncertain whether Class II disposal wells will find wide use in the Marcellus Shale formation. Currently, only four injection wells operate for this purpose in Pennsylvania.

Wastewater injection into the permeable Cambrian sandstones that lie beneath the Marcellus Shale appears feasible. The Cambrian Mt. Simon Sandstone, considered an ideal geologic unit in Ohio for disposal and long-term storage of liquid wastes, is relatively deep, and underlain and overlain by impervious confining layers that prevent migration of injected fluids.[89]

Although underground injection of wastewater may not be practical or economic in all areas across the Marcellus region because of the lack of suitable injection zones, the cost and environmental concerns associated with surface disposal may make Class II injection wells the preferred disposal option for flowback and other wastewater from hydraulic fracturing operations where feasible. This appears to be the trend in other shale areas. In the Fayetteville Shale in Arkansas, trucks have typically collected wastewater and hauled it to disposal wells distant from the producing areas. However, with more intense shale development, the high cost of transporting, treating, and disposing water offsite, injection well use has increased.[90] In the Barnett Shale in Texas, flowback water has been primarily disposed in Class II injection wells.

Both Class II injection and municipal and industrial water treatment facilities are under consideration for the Marcellus Shale, and more than 60 permit applications for such wells are pending in New York for Marcellus Shale development.[91] One firm active in Marcellus Shale development has been disposing of flowback water and produced water using three UIC disposal wells and two commercial water treatment facilities, but reportedly plans to use only disposal wells in the future. Based on leases already held, the firm plans to drill between 13,500 and 17,000 gas wells.[92]

Technical and practical questions regarding the development of the Marcellus Shale remain unanswered. USGS researchers have noted that while drilling and hydraulic fracturing technologies have improved over the past several decades, "the knowledge of how this extraction might affect water resources has not kept pace." [93] Consequently, environmental regulators and gas developers face new challenges and some uncertainties as the Marcellus Shale is developed.

State Water Quality Laws

State laws addressing the quality of surface water and groundwater also appear to apply to Marcellus Shale development. For example, in New York various aspects of such development would require a permit under the State Pollutant Discharge Elimination System (SPDES).[94] SPDES is an "approved," rather than delegated, version of the federal National Pollutant Discharge Elimination System (NPDES) permit program because, while the federal NPDES covers only discharges to surface water, SPDES covers discharges to groundwater also. The SPDES permit requirement could apply to hydraulic fracturing by meeting four conditions:

1. Most importantly, the state must determine that injection will not degrade groundwater;[95]
2. A wastewater treatment plant would likely dispose of the fluids produced from the well, in which case the plant's SPDES permit would apply;
3. SPDES permits would also cover treatment facilities built specially for disposing of flowback water, if there would be discharges into a water body; and
4. Applicable state water-quality standards would control, in part, the permit's discharge limits.[96]

New York State's Environmental Quality Review Act (SEQRA) is also relevant.[97] As with its federal counterpart, the National Environmental Policy Act, a requirement of an environmental impact statement preparation lies at the heart of the statute.[98] New York has been evaluating the environmental impact of the drilling and hydraulic fracturing activities for more than 15 years through a Generic Environmental Impact Statement (GEIS) that sets parameters that apply statewide for SEQRA review of gas well permitting. In February 2009, the state's Bureau of Oil and Gas Regulation, in the Department of Environmental Conservation, released the final scoping document under SEQRA for a Supplemental Generic Environmental Impact Statement (SGEIS) on developing the Marcellus and other gas shale regions in the state using high-volume hydraulic fracturing. On September 30, New York DEC released for comment the draft SGEIS which proposes additional parameters for SEQRA review and focuses on water supply protection and wastewater management as major issues. Until New York finalizes the supplemental GEIS, the state will only accept, but not process, permit applications for gas wells involving horizontal drilling and high-volume hydraulic fracturing.[99]

As another example, West Virginia's NPDES permit program would apply to wastewater treatment plants to which flowback from Marcellus Shale production sites was taken and to treatment facilities built specially for the frac water that discharge into a water body.[100] Applicable state water-quality standards would control the permit's discharge limits, in part.[101] However, this program applies to surface water only, not groundwater, and the state's Groundwater Protection Act exempts "groundwater within geologic formations which are site specific to ... [t]he production ... of ... natural gas.... "[102] The state's underground injection control program would regulate frac water re-injected at a second or subsequent production site.[103]

In addition to state water-quality laws, the interstate Delaware River Basin Commission (36% of whose jurisdictional land area in Pennsylvania and New York overlies the Marcellus

Shale formation) would also impose water quality requirements.[104] The Commission's water quality (and other) requirements are legally separate from those of the affected states—that is, obtaining state approval does not excuse an applicant from seeking Commission approval—though in some cases the two requirements may be substantively identical.

Another interstate-compact-created commission within the Marcellus region, the Susquehanna River Basin Commission, regulates only water quantity, not water quality.

State Water Supply Management

Gas producers must arrange to procure the large volumes of water required for hydraulic fracturing in advance of their drilling and development activity. Generally, water rights and water supply regulation differ among the states. Depending on individual state resources and historic development, states may use one of two water rights doctrines, riparian or prior appropriation, or a hybrid of the two. Under the riparian doctrine, a person who owns land that borders a watercourse has the right to make reasonable use of the water on that land.[105] Traditionally, the only limit to users under the riparian system is the requirement of reasonableness in comparison to other users. Under the prior appropriation doctrine, a person who diverts water from a watercourse (regardless of his location relative thereto) and makes reasonable and beneficial use of the water may acquire a right to use of the water.[106] The prior appropriation system limits users to the quantified amount of water the user secured under a state permitting process with a priority based on the date the state conferred the water right. Because of this priority system, the phrase "first in time, first in right" has sometimes substituted for appropriative rights. Some states have implemented a dual system of water rights, assigning rights under both doctrines.

Generally, states east of the Mississippi River follow a riparian doctrine of water rights, while western states follow the appropriation doctrine.[107] The system of water rights allocation in a particular state with shale resources may affect the development process, particularly in times when shortages in water supply affect the area of shale development. In areas where the Marcellus Shale is located, which are generally riparian states, water rights may not be as big a concern as in other areas of the country with shale development, such as the Barnett Shale in Texas. That is, even in times of shortage, shale development may be able to continue in the Marcellus Shale region because riparian users reduce water usage proportionally and may still receive enough for supply requirements of the development process. On the other hand, appropriative rights users in the Barnett Shale region may not be able to fill their water rights at all if other senior rights take all the water, and thus would have to postpone development. In addition, interstate compacts may apply and affect water supply for shale development processes. In the case of Marcellus Shale development, several interstate compacts are relevant, as discussed below.

New York's SPDES permit program (discussed above) governs water quality only, not water quantity. With a limited exception for pumping water on Long Island,[108] there is no proactive regulatory scheme in New York for extracting water from streams, lakes, groundwater, etc. In the case of drawing water from a public drinking water supplier, however, the state does have limited authority to make sure that the public water supplier stays within its permit terms. Otherwise, however, the state can only *respond* to water flow problems—e.g., if a fish kill occurs, it can prosecute the responsible entity for violating the flow standard that is a component of the state's water quality standards.[109] There is no requirement to notify the state in advance of a water extraction.

West Virginia passed the Water Resources Protection and Management Act in 2003.[110] It requires users of water resources whose withdrawals exceed 750,000 gallons in any given month for one facility to register with the Division of Water and Waste Management in the Department of Environmental Protection.[111] To protect both ground and surface waters, the state proposes to require operators to provide information about the sources of withdrawals, anticipated volumes, and the time of year of withdrawals prior to start-up.[112] State officials believe it is likely that some oil and gas industry operations in the Marcellus Shale region will exceed this threshold and will be required to submit withdrawal information.[113] The goal is to ensure that water withdrawal from ground or surface waters does not exceed volumes beyond what the waters can sustain.[114]

Texas is another relevant example, because of similarities between the Barnett Shale there and the Marcellus Shale. Texas has codified the public trust doctrine regarding ownership of state water resources. That is, water in any of the various waterbodies—including rivers, streams, lakes, etc.—within the state is the property of the state of Texas.[115] Individuals or entities may divert the state's waters for their own use only after acquiring a permit (water right) from the state through its Commission on Environmental Quality.[116] Texas does provide for the possibility of temporary water permits for a period of up to three years, if a temporary permit would not adversely affect senior rights.[117]

Other states apply surface and groundwater regulations similarly, and gas producers using fresh water for drilling and development must comply with state and local administration of water rights.

As for interstate constraints in the Marcellus Shale region and vicinity, the Delaware River Basin Commission[118] and the Susquehanna River Basin Commission[119] impose limits on the quantity of water extracted. In addition, the Great Lakes-St. Lawrence River Basin Water Resources Compact[120] prohibits inter-basin transfers of water.

CONGRESSIONAL INTEREST

Recently proposed legislation would affect natural gas projects in the Marcellus Shale directly and indirectly. The 111th Congress introduced bills to amend the Safe Drinking Water Act to define hydraulic fracture as underground injection for regulatory purposes. Bills to revise the Commodity Exchange Act would place limits on the use of futures markets to hedge the risks associated with natural gas development projects. The recently enacted mandate to increase the use of renewable fuels (ethanol) has an indirect link, as fertilizer produced from natural gas is essential to producing corn — the primary feedstock for ethanol. Recently proposed low-carbon fuel standards may overlook natural gas as a substitute transportation fuel.

Current and planned projects to develop Marcellus Shale gas are apparent across the six-state region that overlies the resource (see **Figure 5**). For example, gas producers have reportedly planned over 2,000 gas wells just in West Virginia, and the state's Oil and Gas Commission estimates that, based on current information, there could be a well on every 40 acres in the state. Throughout the region, this activity is likely to put increasing demands on regulatory agencies—especially state agencies[121]—for necessary licensing, permitting, inspections, and enforcement. As noted by USGS, because of questions related to water supply and wastewater disposal, many state agencies have been cautious about granting permits, and some states have placed *de facto* moratoria on drilling until these issues are resolved.[122] New York, for example, is accepting but not processing permits until it completes a Supplemental Generic Environmental Impact Statement that will impose new environmental review requirements for permits for gas well development using high-volume hydraulic fracturing. The success of planned development activities could depend, in part, on the capacity of regulatory agencies to provide the administrative resources that supporting such plans would require.

The Energy Policy Act of 2005 (P.L. 109-58) in Section 322 (Hydraulic Fracturing) amended the Safe Drinking Water Act (42 U.S.C. 300h(d)(1)) to exclude the fluids or propping agents (other than diesel fuels) used in hydraulic fracturing operations related to oil, gas, or geothermal production activities from the definition of the term "underground injection." Several pending bills address the current exemption of hydraulic fracturing under SDWA:

- H.R. 2300 (introduced May 7, 2009, as the American Energy Innovation Act) would express the sense of Congress that the Safe Drinking Water Act was never intended to regulate natural gas and oil well construction stimulation and that the amendment of SDWA by the Energy Policy Act of 2005 to clarify that the SDWA was not intended to regulate the use of hydraulic fracturing should be maintained.
- H.R. 2766/S. 1215 (introduced June 9, 2009, as the "Fracturing Responsibility and Awareness of Chemicals Act") would amend the SDWA definition of 'underground injection' to include the underground injection of fluids or propping agents used for hydraulic fracturing operations related to oil and gas production activities. The bills would require public disclosure of the chemical constituents (but not the proprietary chemical formulas) used in the fracturing process. Disclosure of a propriety formula to the state, EPA Administrator, or treating physician or nurse would be required in the case of a medical emergency.
- The conference report for the Department of the Interior, Environment, and Related Agencies Appropriations Act, 2010 (H.R. 2996, H.Rept. 111-216) includes a provision urging the EPA, in consultation with appropriate federal, state and interstate agencies, to carry out a study on the relationship between hydraulic fracturing and drinking water.

Many industries make use of various "hedging" strategies to minimize the risk of commodity price increases. A simple hedge involves buying "futures" contracts to lock in prices. For gas exploration and development companies, hedges in effect guarantee the amount of revenue that companies will receive on a future production, thus giving them some financial stability. Two current proposed bills would limit speculation on future commodity prices. Some would argue, however, that in a time of low natural gas prices and scarce capital for financing new energy projects, restrictions on hedging could adversely these companies.

- H.R. 977 (introduced February 11, 2009, as the "Derivatives Markets Transparency and Accountability Act of 2009") would amend the Commodity Exchange Act to bring greater transparency and accountability to commodity markets. The Commodity Futures Trading Commission would establish limits on the amount of positions that a person may hold for future delivery contracts, options on such contracts or on commodities traded.
- S. 447 (introduced February 13, 2009, as the "Prevent Excessive Speculation Act") would amend the Commodity Exchange Act to prevent excessive price speculation with respect to energy commodities.

Congressional concern for energy independence that grew out of the summer 2008 oil price spikes has engendered at least one highly publicized proposal to substitute natural gas for transportation fuel. [123] Among various bills introduced to reduce consumer dependence on fossil fuels, some would establish low-carbon fuel standards. Natural gas (CH_4) has a carbon to hydrogen ratio of 1:4, the lowest of all the fossil fuels, thus leading some to argue that it has a role in newly proposed standards.

- H.R. 1787 (introduced March 30, 2009, and cited as the "Low Carbon Fuel Standard of 2009") would amend the Clean Air Act to establish a low carbon fuel standard for transportation fuels. Based on a lifecycle greenhouse gas emission baseline established for all transportation fuels, transportation fuel providers would have to reduce the annual emissions per unit of energy by 5% after 2023, and 10% after 2030.
- S. 1095 (introduced May 20, 2009, and cited as "America's Low-Carbon Fuel Standard Act of 2009") would amend the Clean Air Act to convert renewable fuel standards into low-carbon fuel standards. Low-carbon fuel would be defined as a transportation fuel that has lifecycle greenhouse gas emissions, equal on an annual average basis to a defined percentage less than baseline lifecycle greenhouse gas emissions. Starting at 20% in 2015, the percentage would increase to 42.5% after 2031. Regulations would insure that increasing volumes of low-carbon fuel would be sold in the United States as transportation fuel; beginning with 10% in 2015, and 32.5% by 2030.

U.S. fertilizer production has a close link to energy availability, particularly natural gas.[124] Natural gas is the key ingredient in producing anhydrous ammonia, used directly as a nitrogen fertilizer, and used as a basic building block for producing most other forms of nitrogen fertilizers (e.g., urea, ammonium nitrate, and nitrogen solutions). Natural gas also serves as a process gas in the manufacture of other nitrogenous fertilizers from anhydrous ammonia. As a result, natural gas accounts for 75% to 90% of costs of production for nitrogen fertilizers. Because fertilizer prices are closely linked to natural gas prices, higher gas prices encourage two potential responses: (1) lower fertilizer application rates on the current farm planting mix; or (2) the planting and production of crops that are less dependent on fertilizer. Among major U.S. field crops, corn uses the most fertilizer according to the U.S. Department of Agriculture's Economic Research Service.[125] Thus, ethanol production costs and thus renewable energy costs are likely to reflect natural gas availability and price. Under the Energy Independence and Security Act of 2007 (P.L. 110-140), gasoline sold in the United

States must contain a minimum volume of renewable fuel. The Renewable Fuel Standard program that results from this requirement will increase the required volume of renewable fuel blended into gasoline from 9 billion gallons in 2008 to 36 billion gallons by 2022.

APPENDIX. GLOSSARY

Bcf: Billion Cubic Feet; a gas measurement equal to 1,000,000,000 cubic feet. See also Mcf, Tcf, Quad.

Btu: British thermal unit; the amount of energy required to heat one pound of water by one degree Fahrenheit.

Frac: Hydraulic fracturing, as adopted by the petroleum industry.

Flowback: The fracture fluids that return to surface after a hydraulic fracture is completed.

Mcf: A natural gas measurement unit for one thousand cubic feet.

MMcf: A natural gas measurement unit for one million cubic feet.

NORM: Natural occurring radioactive material; includes naturally occurring uranium-235 and daughter products such as radium and radon.

Oil-equivalent gas (OEG): The volume of natural gas needed to generate the equivalent amount of heat as a barrel of crude oil. Approximately 6,000 cubic feet of natural gas is equivalent to one barrel of crude oil.

Slickwater: Water-based frac fluid.

Tcf: A natural gas measurement unit for one trillion cubic feet.

Thixotropy: The property of a gel to become fluid when disturbed (as by shaking).

Whipstock: A wedge-shaped piece of metal placed downhole to deflect the drill bit.

ACKNOWLEDGMENTS

Cynthia Brougher, Legislative Attorney

End Notes

[1] "Reserves are those quantities of petroleum, which, by analysis of geoscience and engineering data, can be estimated with reasonable certainty to be commercially recoverable ... " in: Society of Petroleum Engineers: *Petroleum Resources Management System,*2007, p. 3, http://www.spe.org/spe-app/spe/industry/reserves/prms.htm.

[2] See Energy Information Administration, *Natural Gas page*, http://www.eia.doe.gov/oil_gas/natural_gas/info_glance/ natural_gas.html.

[3] *North American Natural Gas Supply Assessment*, Prepared for Clean Skies Foundation, Navigant Consulting, July 4, 2008.

[4] Colorado School of Mines, *Potential Gas Committee reports unprecedented increase in magnitude of U.S. natural gas resource base*, June 18, 2009, http://www.mines.edu/Potential-Gas-Committee-reports-unprecedented-increase-inmagnitude-of-U.S.-natural-gas-resource-base.

[5] U.S. DOE Energy Information Administration, *Natural Gas Navigator*, Natural Gas Summary, June 29, 2009, http://tonto.eia.doe.gov/dnav/ng/ng_sum_lsum_a_EPG0_FGW_mmcf_a.htm.

[6] U.S. DOE Energy Information Administration, *Is U.S. natural gas production increasing?*, June 11, 2008, http://tonto.eia.doe.gov/energy_in_brief/natural_gas_production.cfm.

[7] Navigant Consulting, *Unconventional Gas Supply*, TransCanada Pipelines Customer Meeting, San Francisco, CA, August 26, 2008, http://www.navigantconsulting.com/downloads/knowledge_center/ Unconventional_Natural_Gas_RWelch_Aug_08.pdf.

[8] See Navigant Consulting.

[9] See Navigant Consulting.

[10] StatoilHydro, "Forms strategic alliance with major US gas player," press release, November 11, 2008, http://www.statoilhydro.com/en/NewsAndMedia/News/2008/Pages/Chesapeake.aspx.

[11] Mississippian-aged shale dates back to the Paleozoic era of around 345 to 320 million years ago.

[12] Richard M. Pollastro (Task Leader), Ronald J. Hill, and Thomas A. Albrandt, et al., Assessment of Undiscovered Oil and Gas Resources of the Bend Arch-Fort Worth Basin Province of North-Central Texas and Southwestern Oklahoma, 2003, U.S. Geological Survey, Fact Sheet 2004-3 022, March 2004. http://pubs.usgs.gov/fs/2004/3022/

[13] Richard M. Pollastro, Daniel M. Jarvie, and Ronald J. Hill, et al., "Geologic Framework of the Mississippian Barnett Shale, Barnett-Paleozoic total petroleum system, Bend arch-Fort Worth Basin, Texas," *American Association of Petroleum Geologists*, vol. 91, no. 4 (April 2007), pp. 405-436.

[14] Personal communication with Eric Potter, Jackson School of Geosciences, University of Texas, October 2008.

[15] Pollastro, et al., *Geologic framework of the Mississippian Barnett Shale.*

[16] Energy Information Administration, Office of Oil and Gas, Natural Gas Division , *About U.S. Natural Gas Pipelines*, Major Natural Gas Transportation Corridors, http://www.eia.doe.gov/pub/oil_gas/natural_gas/analysis_publications/ ngpipeline/transcorr.html.

[17] Damien Gaula, *Expansion of the U.S. Natural Gas Pipeline Network: Additions in 2008 and Projects through 2011*, Energy Information Administration, Office of Oil and Gas, September 2009, http://www.eia.doe.gov/pub/oil_gas/ natural_gas/feature_articles/2009/pipelinenetwork/pipelinenetwork.pdf.

[18] James Bene et al., *Northern Trinity/Woodbine Groundwater Availability Model: Assessment of Groundwater Use in the Northern Trinity Aquifer Due to Urban Growth and Barnett Shale Development*, R.W. Harden & Associates, prepared for the Texas Water Development Board, Austin, TX, January 2007, p. 2-2 1, at http://www.twdb.state.tx.us/ RWPG/rpgm_rpts/0604830613_BarnetShale.pdf. Hereafter referred to as R.W. Harden & Associates, 2007.

[19] R.W. Harden & Associates, 2007, p. 2.

[20] Texas Water Development Board, *2007 State Water Plan, Chapter 7, Groundwater Resources*.

[21] In Texas, a minor aquifer is defined as one that supplies large quantities of water in small areas or small quantities of water in large areas; a major aquifer is generally defined as supplying large quantities of water over a large areas of the state. See John B. Ashworth and Janie Hopkins, *Major and Minor Aquifers of Texas*, Texas Water Development Board, Report 345, November 1995, p. 1. The state of Texas recognizes 7 major and 21 minor aquifers, see Texas Water Development Board, *2007 State Water Plan, Chapter 7, Groundwater Resources*, at http://www.twdb.state.tx.us/wrpi/ swp/swp.htm.

[22] Paul D. Ryder, U.S. Geological Survey, *Ground Water Atlas of the United States*, Oklahoma, Texas; HA 730-E, 1996, at http://pubs.usgs.gov/ha/ha730/ch_e/index.html.

[23] R.W. Harden & Associates, 2007, p. 2.

[24] John B. Ashworth and Janie Hopkins, *Major and Minor Aquifers of Texas: Trinity Aquifer*, Texas Water Development Board, Report 345, November 1995.

[25] James Bene et al., *Northern Trinity/Woodbine Groundwater Availability Model: Assessment of Groundwater Use in the Northern Trinity Aquifer Due to Urban Growth and Barnett Shale Development*, R.W. Harden & Associates, p.2.

[26] Other factors could include the thickness and maturity of the shale, the ability to drill horizontally or vertically, the number of well completions per year, availability of water, and regulatory factors, among others.

[27] James Bene et al., *Northern Trinity/Woodbine Groundwater Availability Model: Assessment of Groundwater Use in the Northern Trinity Aquifer Due to Urban Growth and Barnett Shale Development*, R.W. Harden & Associates, p.15.

[28] U.S. Department of Energy Office of Fossil Energy and National Technology Laboratory, *Modern Shale Gas Development in the United States: A Primer*, DE-FG26-04NT 15455, April 2009, http://fossil.energy.gov/programs/ oilgas/publications/naturalgas_general/Shale_Gas_Primer_2009.pdf.

[29] An isopach is a continuous line connecting points of equal thickness.

[30] "Marcellus Shale Play's Vast Potential Creating Stir in Appalachia," by Terry Engelder and Gary Lash, Professors of Geosciences, The American Oil and Gas Reporter, May 2008.

[31] StatoilHydro, "StatoilHydro's acquisition of 32.5% of Chesapeake interest in the Marcellus shale," press release, 2008, http://www.statoilhydro.com/en/NewsAndMedia/News/2008/Downloads/Frequently%20asked%20questions.pdf.

[32] U.S. DOE Energy Information Administration, *Natural Gas Consumption by End Use*, http://tonto.eia.doe.gov/dnav/ ng/ng_cons_sum_dcu_nus_a.htm.

[33] Reported by EIA as 23,047,641 million cubic feet, http://tonto.eia.doe.gov/dnav/ng/ng_cons_sum_dcu_nus_a.htm.

[34] U.S. DOE Energy Information Administration, *Natural Gas Pipelines in the Northeast Region*, http://www.eia.doe.gov/pub/oil_gas/natural_gas/analysis_publications/ngpipeline/northeast.html.

[35] The natural gas received and transported by the major intrastate and interstate mainline transmission systems must meet the quality standards specified by pipeline companies in the "General Terms and Conditions (GTC)" section of their tariffs. These quality standards vary from pipeline to pipeline and are usually a function of a pipeline system's design, its downstream interconnecting pipelines, and its customer base. In general, these standards specify that the natural gas: be within a specific Btu content range (1,035 Btu per cubic foot, +/- 50 Btu); be delivered at a specified hydrocarbon dew point temperature level (below which any vaporized gas liquid in the mix will tend to condense at pipeline pressure); contain no more than trace amounts of elements such as hydrogen sulfide, carbon dioxide, nitrogen, water vapor, and oxygen; and be free of particulate solids and liquid water that could be detrimental to the pipeline or its ancillary operating equipment. U.S. DOE Energy Information Administration, *Natural Gas Processing: The Crucial Link Between Natural Gas Production and Its Transportation to Market,* http://www.eia.doe.gov/pub/oil_gas/ natural_gas/feature_articles/2006/ngprocess/ngprocess.pdf.

[36] Henry Trapp, Jr. and Marilee A. Horn, "U.S. Geological Survey Ground Water Atlas of the United States," HA 730- L (1997).

[37] See New York State, Department of Environmental Conservation, "Primary & Principal Aquifers," at http://www.dec.ny.gov/lands/36119.html.

[38] U.S. Environmental Protection Agency, *Sole Source Aquifer Protection Program*, http://cfpub.epa.gov/safewater/ sourcewater/sourcewater.cfm?action=SSA.

[39] Cattaraugus Creek Basin Aquifer, Cortland Homer-Preble Aquifer System, Clinton Street-Ballpark Valley Aquifer System, and Schenectady/Niskanyuna Aquifers in New York, and Steven Valleys Aquifer in Pennsylvania.

[40] Howard Hughes, Jr. of the Hughes Tool Company developed the modern tri-cone rotary bit. His father, Howard Robert Hughes, Sr. had invented the bit's ancestor, a two-cone rotary bit.

[41] Schlumberger, *Better Turns for Rotary Steerable Drilling: Overview,* http://www.slb.com/content/services/resources/ oilfieldreview/ori002/01 .asp?.

[42] Schlumberger, Inc., *Shale Gas: When Your Gas Reservoir is Unconventional, So is Our Solution.* http://www.slb.com/media/services/solutions/reservoir/shale_gas.pdf.

[43] "Chapter 7 - Fracturing Fluid Chemistry and Proppants," in *Reservoir Stimulation*, ed. Michael J. Economides and Kenneth G. Nolte, 3rd ed. (John Wiley & Sons, LTD, 2000).

[44] 29 C.F.R. §§ 1910 Subpart Z, Toxic and hazardous substances.

[45] Modern Shale Gas Development in the United States: A Primer, pp. 58-59.

[46] U.S. Geological Survey, *Summary of water use in the United States, 2000*, http://ga.water.usgs.gov/edu/ wateruse2000.html.

[47] The term "mineral" refers to economically recoverable resources including natural gas.

[48] This information was provided in a BLM Instructional Memorandum, No. 2003-131. Also see 43 CFR 3104, 43 CFR 3814 and 48 FR 48916 (1983) for further details.

[49] The Natural Gas Leasing Offer Tracking can be found at http://www.pagaslease.com/lease_tracking_2.php.

[50] Natural Gas /Oil Landowner Groups Directory, http://www.pagaslease.com/directory_public.php.

[51] West Virginia Surface Owners' Rights Organization, http://www.wvsoro.org/.

[52] Energy Conservation Law, 23-0301-23-0305, Part 556.

[53] Pennsylvania Budget and Policy Center, *Responsible Growth, Protecting the Public Interest with a Natural Gas Severance Tax*, by Michael Wood and Sharon Ward, April 2009.

[54] Pennsylvania State University, College of Earth and Mineral Sciences, Department of Energy and Mineral Engineering, An *Emerging Giant: Prospects and Economic Impacts of Developing the Marcellus Shale Natural Gas Play*, by Timothy Considine, et al., August 5, 2009.

[55] U.S Department of the Interior, Inventory of Onshore Federal Oil and Natural Gas Resources and Restrictions to Their Development, Phase III Inventory - Onshore United States, 2008, p. 225. http://www.blm.gov/pgdata/etc/medialib/blm/wo/MINERALS__REALTY__AND_RESOURCE_ PROTECTION_/energy/

[56] Daniel J. Soeder and William M. Kappel, *Water Resources and Natural Gas Production from the Marcellus Shale*, U.S. Geological Survey, U.S. Department of the Interior, Fact Sheet 2009-3032, May 2009, http://pubs.usgs.gov/fs/ 2009/3032/pdf/FS2009-3032.pdf. Hereafter, USGS Fact Sheet.

[57] Railroad Commission of Texas, Water Use in the Barnett Shale, July 30, 2008, http://www.rrc.state.tx.us/division/og/ wateruse_barnettshale.html.

[58] Burnett, D.B. and Vavra, C.J., Desalination of Oil Field Brine—Texas A&M Produced Water Treatment. August 2006, http://www.pe.tamu.edu/gpri-new/home/BrineDesal/MembraneWkshpAug06/Burnett8-06.pdf.

[59] USGS Fact Sheet at 4.

[60] These particles, termed normally occurring radioactive materials (NORMS), can be brought to the surface on drilling equipment and in fluids. Subsurface formations may contain low levels of such materials as uranium and thorium and their daughter products, radium 226 and radium 228. On gamma-ray logs, shales can be differentiated from other rocks such as clean sandstones and limestones because shales have higher concentrations of potassium-40-bearing minerals. See Commonwealth of Pennsylvania Department of Conservation and Natural Resources, Pennsylvania Geology, Vol. 38, No. 1, p. 5, Spring 2008, http://www.dcnr.state.pa.us/topogeo/pub/pageolmag/pdfs/v38n1.pdf.

[61] USGS Fact Sheet at 4.

[62] In the Barnett Shale area, most of the water is reinjected for disposal.

[63] 33 U.S. C. § 1313.

[64] Under CWA Section 301, it is illegal to discharge pollutants into the nation's waters except in compliance with substantive and procedural provisions of the law, which include obtaining a discharge permit. 33 U.S.C. § 1311.

[65] "Minimal Impact on Total Dissolved Solids Found in Monongahela River Last Fall," Natural Gas, Tapping Pennsylvania's Potential, May 20, 2009, http://www.pamarcellus.com/news.php.

[66] Marcellus Shale Wastewater Partnership, "Permitting Strategy for High Total Dissolved Solids (TDS) Wastewater Discharges," April 11, 2009, http://www.depweb.state.pa.us/watersupply/cwp/view.asp?a=1260&Q=545730& watersupplyNav=|.

[67] Legere, Laura, "How to handle wastewater big challenge in gas drilling," The Citizens' Voice, August 25, 2008.

[68] State of West Virginia, Department of Environmental Protection, Office of Oil and Gas, "Industry Guidance, Gas Well Drilling/Completion, Large Water Volume Fracture Treatments (Draft)," March 13, 2009, p. 3, http://www.wvsoro.org/curent_events/marcellus/Marcellus_Guidance_Draft.pdf.

[69] Ibid. at 3-4. Reportedly, one company with wells in the Marcellus shale in West Virginia has its hydraulic fracturing wastewater trucked to an out-of-state commercial facility that treats the water and then injects in into depleted oil and gas reservoirs. (Kasey, Pam, *New Drilling Efforts Raise Questions*, The State Journal, August 14, 2008.)

[70] West Virginia Surface Owners' Rights Organization, "Proposed Changes to Oil & Gas Rules, Marcellus Guidance Document," http://www.wvsoro.org/curent_events/.

[71] Clean Water Act section 402(p); 33 U.S.C. § 1342(p).

[72] 33 U.S.C. § 1342(l)(2); 33 U.S.C. §1362(24).

[73] The Pennsylvania permit is called an Earth Disturbance Permit (ESCGP-1).

[74] EPA retains the authority to regulate the use of diesel fuel for the purpose of hydraulic fracturing if the agency considers such regulation necessary to protect underground sources of drinking water.

[75] SDWA § 1414(e); 42 U.S.C. § 300g-3(e).

[76] Application, construction, operating, monitoring, and reporting requirements for Class II wells are found in 40 CFR 144 and 146.

[77] To receive primacy, a state, territory, or Indian tribe must demonstrate to EPA that its UIC program is at least as stringent as the federal standards; the state, territory, or tribal UIC requirements may be more stringent than the federal requirements. For Class II wells, states must demonstrate that their programs are effective in preventing pollution of underground sources of drinking water (USDWs). Requirements for state UIC programs are established in 40 CFR §§ 144-147.

[78] SDWA § 142 1(b)(2)

[79] P.L. 109-58, Section 322, amended SDWA section 1421(d).

[80] SDWA §§ 1421 - 1426; 42 U.S.C. §§ 300h - 300h-5. The Safe Drinking Water Act of 1974 (P.L. 93-523) authorized the UIC program at EPA.

[81] Section 1421(d)(2) states:

underground injection endangers drinking water sources if such injection may result in the presence in underground water which supplies or can reasonably be expected to supply any public water system of any contaminant, and if the presence of such contaminant may result in such system's not complying with any national primary drinking water regulation or may otherwise adversely affect the health of persons.

[82] P.L. 93-523, SDWA §1421 (42 U.S.C. § 300h).

[83] 42 U.S.C. § 300i.

[84] Other Class II wells include oil and natural gas storage wells and enhanced oil and gas recovery wells.

[85] U.S. Environmental Protection Agency, Underground Injection Control Program. Oil and Gas Injection Wells: Class II, http://www.epa.gov/safewater/uic/wells_class2.html.

[86] The largest subclass of Class II wells are enhanced recovery wells. These wells are used to inject various substances (including brine, water, steam, polymers, and carbon dioxide) into hydrocarbon-bearing formations to recover primarily oil, but also natural gas, that remains in previously produced areas. Class II enhanced recovery wells are regulated under the UIC program, except for their use in hydraulic fracturing operations.

[87] 25 PA Code § 78.51. Protection of water supplies.

[88] 25 PA Code § 78.18. Disposal and enhanced recovery wells.

[89] Ohio Department of Natural Resources, Division of Geological Survey, *The Geology of Ohio—The Cambrian*, GeoFacts No. 20, May 1998.

[90] University of Arkansas and Argonne National Lab, Reducing the Environmental Impact of Natural Gas Development, http://lingo.cast.uark.edu/LINGOPUBLIC/.

[91] J. Daniel Arthur, et al, Evaluating the Environmental Implications of Hydraulic Fracturing in Shale Gas Reservoirs, 2008, available at http://www.all-llc.com.

[92] StatoilHydro, Frequently Asked Questions: *StatoilHydro 's Acquisition of 32.5% of Chesapeake Interest in the Marcellus Shale*, 2008. http://www.statoilhydro.com/en/NewsAndMedia/News/2008/Downloads/Frequently%20asked%20questions.pdf

[93] USGS Fact Sheet at 5.

[94] N.Y. Envtl. Cons. Law § 17-0505.

[95] N.Y. Code of Rules and Regulations (Conservation) § 750-1.5(a)(6).

[96] N.Y. Envtl. Cons. Law § 17-0501.

[97] N.Y. Envtl. Cons. Law §§ 8-0101 – 8-0117.

[98] Id. at § 8-0109.

[99] In the announcement of the draft supplemental GEIS, the NY Department of Environmental Conservation noted that,

While the process of preparing the Supplemental GEIS is ongoing, any entity that applies for a drilling permit for horizontal drilling in the Marcellus Shale and opts to proceed with its permit application will be required to undertake an individual, site-specific environmental review. That review must take into account the same issues being considered in the Supplemental GEIS process and must be consistent with the requirements of the State Environmental Quality Review Act and the state Environmental Conservation Law.

See http://www.dec.ny.gov/energy/46288.html.

[100] W. Va. Code Ann. § 22-1 1-4(a)(16). See regulations at W. Va. Code of State Rules tit. 47, ser. 10.

[101] W. Va. Code of State Rules tit. 47, ser. 2.

[102] W. Va. Code Ann. § 22-12-5(i).

[103] W. Va. Code Ann. § 22-1 1-8(b)(7). See regulations at W. Va. Code of State Rules tit. 47, ser. 13.

[104] The compact creating the Delaware River Basin Commission was ratified by Congress: P.L. 87-328, 75 Stat. 688. Section 3.8 of the Compact states: "No project having a substantial effect on the water resources of the basin shall hereafter be undertaken by any person, corporation, or government authority unless it shall have been first submitted to and approved by the commission.... " Section 2.3.5 B of the Delaware River Basin Comm'n Administrative Manual (Rules of Practice and Procedure) lists 18 types of projects that must be submitted to the Commission, including withdrawal of groundwater and discharge of pollutants into surface or ground waters of the Basin. Codified at 18 C.F.R. § 401.35(b).

[105] *See generally* A. Dan Tarlock, Law of Water Rights and Resources, ch. 3 "Common Law of Riparian Rights."

[106] *See generally id.* at ch. 5, "Prior Appropriation Doctrine."

[107] The distinction between these doctrines arises primarily from the historic availability of water geographically. In the generally wetter, eastern riparian states, water users share the water resources because water availability historically did not pose a problem to settlement and development. In the drier, western states that experience regular water shortages, the prior appropriation system provides a definitive hierarchy that allows users to acquire well-defined rights to water as a limited resource that requires planning to avoid scarcity.

[108] N.Y. Code of Rules and Regulations (Conservation) § 602.1.

[109] N.Y. Code of Rules and Regulations (Conservation) § 703.2. For certain classes of water bodies, the flow standard prohibits any "alteration that will impair the waters for their best usages."

[110] W. Va. Code Ann. § 22-26.

[111] *Id.* at § 22-26-3.

[112] *See id.*

[113] *See* State of West Virginia, Department of Environmental Protection, Office of Oil and Gas, "Industry Guidance, Gas Well Drilling/Completion, Large Water Volume Fracture Treatments (Draft)," March 13, 2009, pp, 1-2, http://www.wvsoro.org/curent_events/marcellus/Marcellus_Guidance_Draft.pdf.

[114] *Id.*

[115] Tex. Water Code § 11.021.

[116] *See* Tex. Water Code § 11.022.

[117] Tex. Water Code § 11.138.

[118] Congress ratified the compact creating the Delaware River Basin Commission in Pub. Law 87-328, 75 Stat. 688.

[119] Congress ratified the compact creating the Susquehanna River Basin Commission in Pub. Law 91-575, 84 Stat. 1509.

[120] For the text of the compact, see http://www.cglg.org/projects/water/docs/12-13-05/Great_Lakes-St_Lawrence_River_Basin_Water_Resources_Compact.pdf. Congress ratified the compact in P.L. 110-342, 122 Stat. 3739.

[121] As discussed above, states have primary responsibility for many regulatory programs, several interstate commissions also are involved, as are federal agencies such as EPA that implements the UIC program and issues UIC permits in New York and Pennsylvania.

[122] USGS Fact Sheet at 5.

[123] See PickensPlan, http://www.pickensplan.com/act/.

[124] For further information refer to CRS Report RL32677, *Energy Use in Agriculture: Background and Issues*, by Randy Schnepf.

[125] Wen-yuan Huang, William McBride, and Utpal Vasavada, *Recent Volatility in U.S. Fertilizer Prices*, USDA Economic Research Service, March 2009, http://www.ers.usda.gov/AmberWaves/March09/Features/FertilizerPrices.htm.

In: Shale Gas Development
Editor: Katelyn M. Nash pp.151-161

ISBN: 978-1-61668-545-4

Chapter 3

WATER RESOURCES AND NATURAL GAS PRODUCTION FROM THE MARCELLUS SHALE[#]

Daniel J. Soeder[1*] ***and William M. Kappel***[2**]
[1]U.S. Department of the Interior
[2]U.S. Geological Survey

INTRODUCTION

The Marcellus Shale is a sedimentary rock formation deposited over 350 million years ago in a shallow inland sea located in the eastern United States where the present-day Appalachian Mountains now stand (de Witt and others, 1993). This shale contains significant quantities of natural gas. New developments in drilling technology, along with higher wellhead prices, have made the Marcellus Shale an important natural gas resource.

The Marcellus Shale extends from southern New York across Pennsylvania, and into western Maryland, West Virginia, and eastern Ohio (figure 1). The production of commercial quantities of gas from this shale requires large volumes of water to drill and hydraulically fracture the rock. This water must be recovered from the well and disposed of before the gas can flow. Concerns about the availability of water supplies needed for gas production, and questions about wastewater disposal have been raised by water-resource agencies and citizens throughout the Marcellus Shale gas development region. This Fact Sheet explains the basics of Marcellus Shale gas production, with the intent of helping the reader better understand the framework of the water-resource questions and concerns.

[#] This is an edited, reformatted and augmented version of a U. S. Department of the Interior publication dated May 2009.

[*] MD-DE-DC Water Science Center, 5522 Research Park Drive, Baltimore, MD 21228
[**] New York Water Science Center, 30 Brown Road, Ithaca, NY 14850

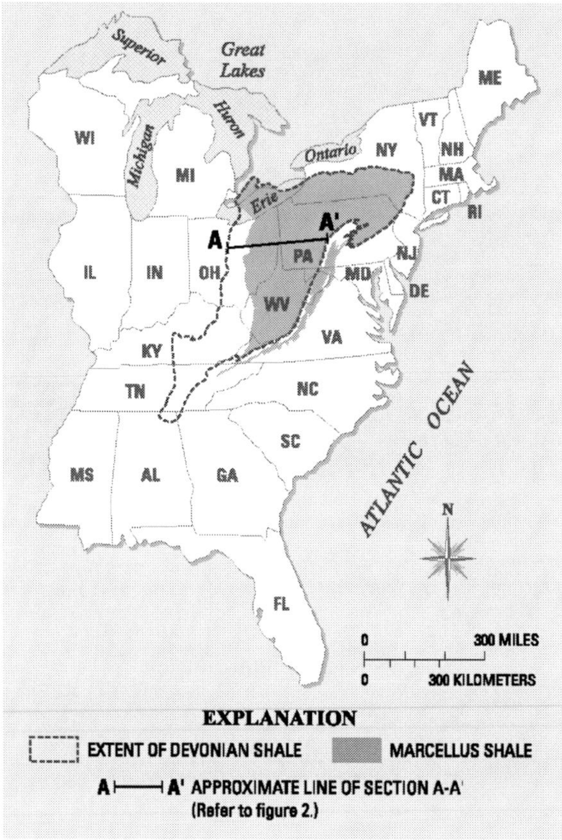

Figure 1. Distribution of the Marcellus Shale (modified from Milici and Swezey, 2006)

What is the Marcellus Shale?

The Marcellus Shale forms the bottom or basal part of a thick sequence of Devonian age, sedimentary rocks in the Appalachian Basin. This sediment was deposited by an ancient river delta, the remains of which now form the Catskill Mountains in New York (Schwietering, 1979). The basin floor subsided under the weight of the sediment, resulting in a wedge-shaped deposit (figure 2) that is thicker in the east and thins to the west. The eastern, thicker part of the sediment wedge is composed of sandstone, siltstone, and shale (Potter and others, 1980), whereas the thinner sediments to the west consist of finer-grained, organic-rich black shale, interbedded with organic-lean gray shale. The Marcellus Shale was deposited as an organic-rich mud across the Appalachian Basin before the influx of the majority of the younger Devonian sediments, and was buried beneath them.

Why Is the Marcellus Shale An Important Gas Resource?

Organic matter deposited with the Marcellus Shale was compressed and heated deep within the Earth over geologic time, forming hydrocarbons, including natural gas. The gas occurs in fractures, in the pore spaces between individual mineral grains, and is chemically adsorbed onto organic matter within the shale (Soeder, 1988). To produce commercial amounts of natural gas from such fine-grained rock, higher permeability flowpaths must be intercepted or created in the formation. This is generally done using a technique called hydraulic fracturing or a "*hydrofrac*," where water under high pressure forms fractures in the rock, which are propped open by sand or other materials to provide pathways for gas to move to the well. Petroleum engineers refer to this fracturing process as "*stimulation*."

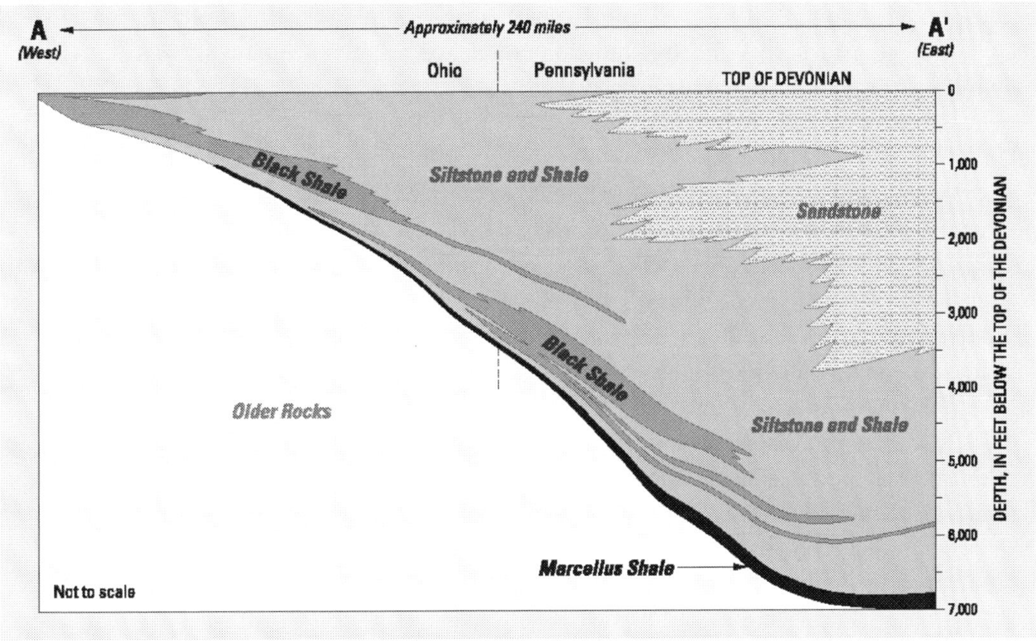

Figure 2. West to east line of section A-A' of Middle and Upper Devonian rocks in the Appalachian Basin. The Marcellus Shale is the lowest unit in the sequence (modified from Potter and others, 1980)

Figure 3. Marcellus Shale drill core from West Virginia, 3.5 inches in diameter, containing a calcite-filled vertical natural fracture. Photograph by Daniel Soeder, USGS

From the mid-1970s to early 1980s, the U.S. Department of Energy (DOE) funded the Eastern Gas Shales Project (EGSP) to develop new technology in partnership with industry that would advance the commercial development of Devonian shale gas (Schrider and Wise, 1980). Goals of the project included assessing the size of the resource, estimating recoverable gas, and determining the most effective technology for gas extraction. The EGSP shale stimulation experiments tested a wide variety of hydrofracs and other techniques. Results were somewhat uneven, and DOE concluded that stimulation alone was generally insufficient to achieve commercial shale gas production (Horton, 1982). It was suggested that better success could be obtained by targeting specific formations in specific locations. The EGSP results did indicate that if the hydraulic fractures were able to intercept sets of existing, natural fractures within the shale (figure 3), a network of flowpaths could be created.

In the mid-1980s, the Institute of Gas Technology (IGT) in Chicago performed some laboratory analyses on EGSP shale samples (Soeder, 1988). Published IGT lab measurements found that a "*gas-in-place*" value for the Marcellus Shale at pressures representative of production depths may be as high as 26.5 standard cubic feet of gas per cubic foot of rock (Soeder, 1988). This greatly exceeded earlier gas-in-place estimates for Devonian shale by the National Petroleum Council (1980), which ranged from 0.1 to 0.6 standard cubic feet of gas per cubic foot of rock. Although IGT analyzed only one sample of Marcellus Shale, the large volumes of gas now being produced from this formation substantiate the early discovery of significant gas reserves.

In 2008, two professors at Pennsylvania State University and the State University of New York (SUNY) Fredonia estimated that about 50 TCF (trillion cubic feet) of recoverable natural gas could be extracted from the Marcellus Shale (Engelder and Lash, 2008). In November 2008, on the basis of production information from Chesapeake Energy Corporation, the estimate of recoverable gas from the Marcellus Shale was raised to more than 363 TCF (Esch, 2008). The United States uses about 23 TCF of natural gas per year (U.S. Energy Information Administration, 2009), so the Marcellus gas resource may be large enough to supply the needs of the entire Nation for roughly 15 years at the current rates of consumption.

Why Is the Marcellus Shale Being Developed Now?

Low prices for natural gas, and ineffective production technology did little to spark interest in Devonian shale gas in the 1990s, despite the publication of the IGT Marcellus gas-in-place estimates. Two factors working together have promoted the current high levels of interest in the Marcellus Shale. First, wellhead prices for gas have risen from values of less than $2.00 per MCF (thousand cubic feet) in the 1980s (figure 4) to a peak of $10.82 per MCF in the summer of 2008 (U.S. Energy Information Administration, 2009). Although prices had declined to $5.15 per MCF in January 2009 due to the economic down-turn, they are still substantially higher than a decade earlier.

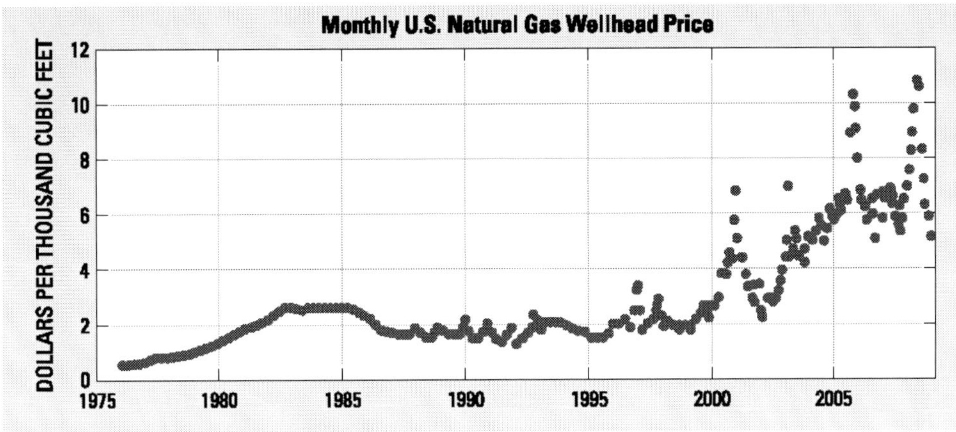

Source U.S. Energy Information Administration, 2009.

Figure 4. Wellhead price of natural gas since the mid-1970s to January 2009

The second factor that spurred interest in the development of the Marcellus Shale is a new application of an existing drilling technology known as *directional drilling*, which involves steering a downhole drill bit in a direction other than vertical. An initially vertical drillhole is slowly turned 90 degrees to penetrate long horizontal distances, sometimes over a mile, through the Marcellus Shale bedrock. Hydraulic fractures are then created into the rock at intervals from the horizontal section of the borehole, allowing a substantial number of high-permeability pathways to contact a large volume of rock (figure 5). According to Range Resources (2008), one of the first major horizontal drillers of Marcellus Shale, these wells typically produce gas at a rate of about 4 MMCF (million cubic feet) per day. Over its lifetime, each horizontal well on an 80-acre surface spacing can be expected to produce a total of about 2.5 BCF (billion cubic feet) of gas at an estimated production cost of $1.00 per MCF (Chernoff, 2008).

WHAT ARE THE WATER-RESOURCE CONCERNS ABOUT DEVELOPING NATURAL GAS WELLS IN THE MARCELLUS SHALE?

Substantial amounts of water are required for the drilling and stimulation of a Marcellus Shale gas well. Fluids recovered from the well, including the liquids used for the hydrofrac, and any produced formation brines, must be treated and disposed of properly. Three important water-resource concerns related to Marcellus Shale gas production are:

- supplying water for well construction without impacting local water resources,
- avoiding degradation of small watersheds and streams as substantial amounts of heavy equipment and supplies are moved around on rural roads, and
- determining the proper methods for the safe disposal of the large quantities of potentially contaminated fluids recovered from the wells.

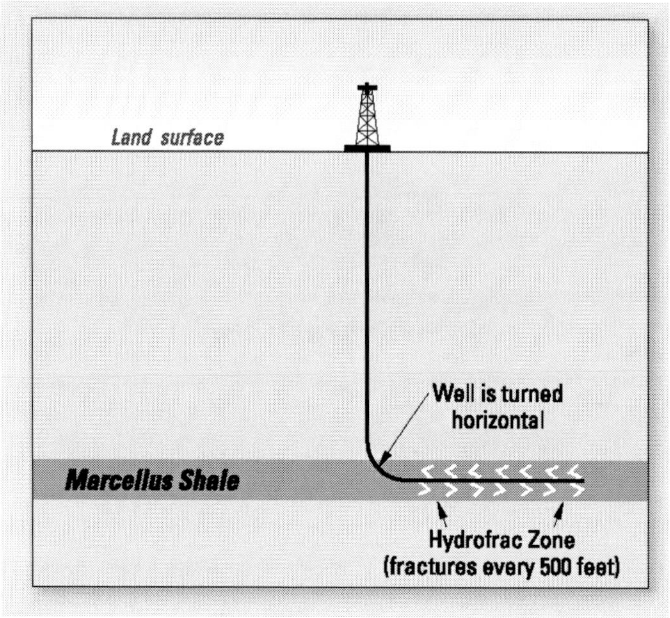

Figure 5. Combination of directional drilling and hydraulic fracturing technology used for gas production from the Marcellus Shale in the Appalachian Basin (modified from http://geology.com/articles/marcellus-shale.shtml)

Figure 6. A hydraulic fracturing stimulation in 2007 on a Marcellus Shale gas well showing the amount of equipment involved

These concerns are discussed in more detail in the following sections.

Water Supply

Drilling requires large amounts of water to create a circulating mud that cools the bit and carries the rock cuttings out of the borehole. After drilling, the shale formation is then stimulated by hydraulic fracturing, which may require up to 3 million gallons of water per treatment (Harper, 2008). Many regional and local water management agencies are concerned about where such large volumes of water will be obtained, and what the possible consequences might be for local water supplies. Under drought conditions, or in locations with already stressed water supplies, obtaining the millions of gallons needed for a shale gas well could be problematic. Drillers could face substantial transportation costs if the water has to be trucked in from great distances.

Similar shale gas operations in the Barnett Shale of Texas have obtained hydrofrac water largely from ground-water sources (Byrd, 2007). Water-supply concerns over the Barnett Shale drilling have been brought up in the past (see, for example, Francis, 2007). Texas State and County agencies now closely monitor volumes of water used during drilling, and a consortium of Barnett Shale drilling companies have developed best management practices for water conservation, with the goal of keeping the pace of drilling and production activities within the bounds of sustainable water use. Similar steps have been discussed in Marcellus Shale gas production areas, but not yet fully implemented.

Transporting Fluids and Supplies

Large hydrofrac treatments often involve moving large amounts of equipment, vehicles, and supplies into remote areas (figure 6). Transporting all of this to drill sites over rural Appalachian Mountain roads could potentially cause erosion, and threaten local small watersheds with sediment. Drill pad and pipeline construction also have the potential to cause similar problems. Of equal concern is the possibility for spills or leaks into water bodies as the fluids and chemical additives are transported and handled. Little is known about how a Marcellus Shale drilling "boom" might adversely affect the land, streams, and available water supplies in the Appalachian Basin. Even under current Marcellus gas production levels, complaints of rural road damage and traffic disrup-tion from drilling equipment have been received, indicating that this could be a significant problem if carried out across thousands of active drill sites.

Wastewater Disposal

For gas to flow out of the shale, nearly all of the water injected into the well during the hydrofrac treatment must be recovered and disposed of. In addition to the problem of dealing with large bulk volumes of liquid waste, contaminants in the water may complicate wastewater treatment. Whereas the percentage of chemical additives in a typical hydrofrac fluid is commonly less than 0.5 percent by volume, the quantity of fluid used in these hydrofracs is so large that the additives in a three million gallon hydrofrac job, for example, would result in about 15,000 gallons of chemicals in the waste.

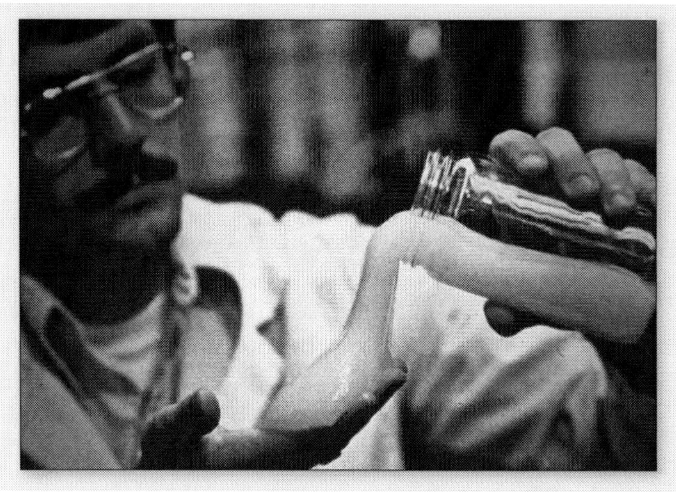

Figure 7. Example of a gel used in hydrofracturing to carry proppant into a fracture. Photograph by Daniel Soeder, USGS.

Hydrofrac fluids are often treated with proprietary chemicals to increase the viscosity to a gel-like consistency that enables the transport of a proppant, usually sand, into the fracture to keep it open after the pressure is released (figure 7). The viscosity of these fluids then breaks down quickly after completion of the hydrofrac, so they can be easily removed from the ground. The chemical formulations required to achieve this are highly researched and closely guarded, and finding out exactly what is in these fluids may present a challenge. The data publicly available on Marcellus Shale hydrofrac treatments indicate that a *slickwater frac* works best on this formation (Harper, 2008). These types of hydrofracs employ linear gels and friction reducers in the water, and utilize only small amounts of proppant, relying instead on fracture surface roughness to hold it open (Rushing and Sullivan, 2007). The potential problems for local wastewater treatment facilities caused by proprietary chemical additives in hydrofrac fluid are unclear.

Along with the introduced chemicals, hydrofrac water is in close contact with the rock during the course of the stimulation treatment, and when recovered may contain a variety of formation materials, including brines, heavy metals, radionuclides, and organics that can make wastewater treatment difficult and expensive. The formation brines often contain relatively high concentrations of sodium, chloride, bromide, and other inorganic constituents, such as arsenic, barium, other heavy metals, and radionuclides that significantly exceed drinking-water standards (Harper, 2008).

The current disposal practice for Marcellus Shale liquids in Pennsylvania requires processing them through wastewater treatment plants, but the effectiveness of standard wastewater treatments on these fluids is not well understood. In particular, salts and other dissolved solids in brines are not usually removed successfully by wastewater treatment, and reports of high salinity in some Appalachian rivers have been linked to the disposal of Marcellus Shale brines (Water and Wastes Digest, 2008). Another disposal option involves re-injecting the hydrofrac fluids back into the ground at a shallower depth. This is a common practice in the Barnett Shale production area of Texas, and has been utilized for some Marcellus wells drilled in West Virginia (Kasey, 2008). Concerns in Appalachian States about the possible contamination of drinking-water supply aquifers has limited the practice of re-injecting Marcellus fluids, however. Another option might be to inject the waste fluid into deeper formations below the Marcellus Shale that are not used as aquifers, such as the Oriskany or Potsdam Sandstones. A third disposal process used in Texas places the wastewater into an open tank to evaporate. The solids that remain behind are then disposed of as dry waste. Although this may be an effective technique in the deserts of the American Southwest, its usefulness in the humid climate of the Appalachians is questionable. A systematic study of the options for Marcellus Shale waste fluid treatment, disposal, or recycling could help to determine the best available procedures.

SUMMARY

Natural gas is an abundant, domestic energy resource that burns cleanly, and emits the lowest amount of carbon dioxide per calorie of any fossil fuel. The Marcellus Shale and other natural gas resources in the United States are important components of a national energy program that seeks both greater energy independence and greener sources of energy. Marcellus gas development has begun in the northern Appalachian Basin, with significant lease holdings throughout Pennsylvania, West Virginia, southern New York, western Maryland, and eastern Ohio. Because of questions related to water supply and wastewater disposal, however, many state agencies have been cautious about granting permits, and some states have placed moratoriums on drilling until these issues are resolved. At the same time, gas companies, drillers, and landowners are eager to move forward and develop the resource.

While the technology of drilling directional boreholes, and the use of sophisticated hydraulic fracturing processes to extract gas resources from tight rock have improved over the past few decades, the knowledge of how this extraction might affect water resources has not kept pace. Agencies that manage and protect water resources could benefit from a better understanding of the impacts that drilling and stimulating Marcellus Shale wells might have on water supplies, and a clearer idea of the options for wastewater disposal.

REFERENCES CITED

Byrd, C. L. (2007). Updated evaluation for the Central Texas Trinity Aquifer Priority Groundwater Management Area, *Priority Groundwater Management Area File Report: Austin*, TX, Texas Commission on Environmental Quality, Water Rights Permitting and Availability Section, Water Supply Division, 160 p.

Chernoff, Harry (2008). *Investing in the Marcellus Shale, article on Seeking Alpha (investing)* website, accessed March 17, 2009 at http://seekingalpha.com/article/68716-investing-in-the-marcellus-shale.

de Witt, Wallace, Jr., Roen, J. B. & Wallace, L. G. (1993). *Stratigraphy of Devonian black shales and associated rocks in the Appalachian basin*, in Roen, J.B., and Kepferle, R.C., eds., Petroleum geology of the Devonian and Mississippian black shale of eastern North America: U.S. Geological Survey Bulletin 1909–B, p. B1–B57.

Engelder, Terry & Lash, Gary (2008). *Unconventional natural gas reservoir could boost U.S. supply, Penn State Live*, accessed January 10, 2009 at http://live.psu.edu/story/28116.

Esch, Mary, (2008). *Estimated gas yield from Marcellus shale goes up*: Albany, NY, Associated Press, November 4, 2008, accessed March 17, 2009 at http://www.ibtimes.com/ articles/20081104/estimated-gasyield-from-marcellus-shale-goes-up. htm.

Francis, R. (2007). *Data on water use in the Barnett Shale begins to surface: Fort Worth*, TX, Fort Worth Business Press, March 12, 2007, accessed March 16, 2009 at http://www.fwbusinesspress.com/display.php?id=5838.

Harper, J. A. (2008). The Marcellus Shale — an old "new" gas reservoir, *in Pennsylvania Geology: Pennsylvania Department of Conservation and Natural Resources*, v. *38*, no. 1, 20 p.

Horton, A. I. (1982). 95 stimulations in 63 wells — DOE reports on comparative analysis of stimulation strategy in eastern gas shales: Columbus, OH, Northeast Oil Reporter, p. 63–73, February 1982.

Kasey, Pam (2008). New drilling efforts raise questions: Charleston, WV, The State Journal, August 14, 2008, accessed March 16, 2009 at http:// statejournal.com/story.cfm?func=viewstory&storyid= 42542.

Milici, R. C. & Swezey, C. S. (2006). *Assessment of Appalachian Basin oil and gas resources: Devonian shale — Middle and Upper Paleozoic total petroleum system*: U.S. Geological Survey Open-File Report 2006–1237, 70 p., accessed April 22, 2009 at http://pubs.usgs.gov/of/2006/1237/.

National Petroleum Council, (1980). *Unconventional gas sources*, Volume III: Devonian shale: Report prepared for the U.S. Secretary of Energy, June 1980, National Petroleum Council, Washington, D.C. 20006, 252 p., accessed March 16, 2009 at http:// www.npc.org/.

Potter, P. E., Maynard, B. & Pryor, W. A. (1980). *Final report of special geological, geochemical, and petrological studies of the Devonian shales of the Appalachian Basin*, prepared for U.S. Department of Energy under contract DE–AC21–76MC05201: Cincinnati, OH, University of Cincinnati, January 1980, [variously paged].

Range Resources (2008). *Range provides Marcellus Shale update, corporate press release: Fort Worth*, TX, Range Resources Corporation, July 14, 2008, accessed March 16, 2009 at http://www.rangeresources. com/PressReleases. asp.

Rushing, J. A. & Sullivan, R. B. (2007). *Improved water frac increases production*: Houston, TX, E&P Magazine, Hart Energy Publishing, September 2007, accessed January 28, 2008 at http://www.epmag.com/ archives/features/661.htm.

Schrider, L. A. & Wise, R. L. (1980). Potential new sources of natural gas: J*ournal of Petroleum Technology*, v. *32*, no. 4, p. 703–7 16, DOI 10.2118/7628–PA.

Schwietering, J. F. (1979). Devonian shales of Ohio and their eastern and southern equivalents: Morgantown, WV, Report METC/CR–79/2, prepared for U.S. Department of Energy under contract EY–76–C–05–5 199, West Virginia Geologic and Economic Survey, January 1979.

Soeder, D. J. (1988). Porosity and permeability of eastern Devonian gas shale: Society of Petroleum Engineers Formation Evaluation, *Society of Petroleum Engineers*, v. *3*, no. 2, p. 116–124, DOI 10.2118/15213–PA.

U.S. Energy Information Administration (2009). U.S. natural gas total consumption, accessed April 15, 2009 at http://www.eia.doe.gov/ oil_gas/natural_gas/info_glance/ natural_gas. html.

Water and Wastes Digest (2008). *Pennsylvania DEP investigates elevated TDS in Monongahela River*, accessed April 10, 2009 at http:// www.wwdmag.com/Pennsylvania-DEP-Investigates-Elevated- TDS-in-Monongahela-River- NewsPiece1 6950.

INDEX

A

B

C

D

E

F

G

H

I

N

O

P

Q

R

S

T

U

V

W

X

Y